Basic Cloning Techniques

Basic Cloning Techniques
A MANUAL OF
EXPERIMENTAL PROCEDURES

EDITED BY

R. H. PRITCHARD BSc, PhD
Professor and Head of
Department of Genetics
School of Biological Sciences
University of Leicester

AND

I. B. HOLLAND BSc, PhD
Professor of Genetics
School of Biological Sciences
University of Leicester

Blackwell Scientific Publications
OXFORD LONDON EDINBURGH
BOSTON PALO ALTO MELBOURNE

Ref
QH
442.2
.B37
1985

© 1985 by
Blackwell Scientific Publications
Editorial offices:
Osney Mead, Oxford, OX2 0EL
8 John Street, London, WC1N 2ES
23 Ainslie Place, Edinburgh, EH3 6AJ
52 Beacon Street, Boston, Massachusetts 02108, USA
667 Lytton Avenue, Palo Alto, California 94301, USA
107 Barry Street, Carlton, Victoria 3053, Australia

All rights reserved. No part of this publication may be
reproduced, stored in a retrieval system, or transmitted,
in any form or by any means, electronic, mechanical,
photocopying, recording or otherwise without the prior
permission of the copyright owner

First published 1985

Photoset by Enset (Photosetting), Midsomer Norton,
Bath, Avon and printed and bound in Great Britain
by Billing & Sons Limited, Worcester

DISTRIBUTORS

USA and Canada
Blackwell Scientific Publications Inc
PO Box 50009, Palo Alto
California 94303

Australia
Blackwell Scientific Publications (Australia) Pty Ltd
107 Barry Street, Carlton, Victoria 3053

British Library
Cataloguing in Publication Data

Basic cloning techniques a manual of experimental
procedures.
1. Cloning
I. Pritchard, R.H.   II. Holland, I.B.
575.1   QH442.2

ISBN 0-632-01032-0

Contents

Contributors, viii
Preface, ix

Section A **Mammalian cDNA cloning**

Introduction, 3

Section A.1 *Isolation of mRNA and translation* in vitro, 6
    A.1 Isolation of mRNA, 8
    A.2 Purification of polyadenylated mRNA by affinity chromatography on oligo-dT cellulose, 10
    A.3 *In vitro* translation of mRNA, 12

Section A.2 *Synthesis of double-strand cDNA and the construction of hybrid pAT153/cDNA plasmids by the homopolymeric tailing method,* 14
    A.4 First-strand cDNA synthesis, 15
    A.5 Second-strand cDNA synthesis, 20
    A.6 S1 nuclease treatment of double-stranded cDNA, 24
    A.7 Homopolymeric tailing of double-stranded cDNA, 27
    A.8 Annealing of oligo dC tailed double-stranded cDNA with oligo dG tailed plasmid DNA, 32

Section A.3 *Transformation of* Escherichia coli *cells with cDNA-containing recombinant plasmid molecules and analysis of cDNA clone-bank,* 34
    A.9 Transformation of competent *E. coli* cells by plasmid DNA, 35
    A.10 Analysis of recombinant plasmid transformants, 39
    A.11 Analysis of the plasmids present in recombinant clones, 46

Appendixes
    A.I Preparation and use of micrococcal nuclease-treated reticulocyte lysates, 51
    A.II Alternative method for second-strand cDNA synthesis, 54
    A.III A procedure for characterizing preparations of S1 nuclease, 55
    A.IV Analysis of products of cDNA synthesis by denaturing polyacrylamide gel electrophoresis, 57
    A.V Additional notes and trouble shooting for Experiment A.9, 59
    A.VI Single colony SDS lysate, 62
    A.VIII Preparation of plasmid DNA, 63

Section B   **Analysis of DNA and RNA**

Introduction, 71

- B.1   Construction of plasmid restriction maps, 72
- B.2   Detection of mammalian β-globin genes in total genomic DNA by filter hybridization with cloned β-globin cDNA, 80
- B.3   Spot hybridizations of animal DNAs, 91
- B.4   Northern blotting, 93

Appendixes

- B.I   Additional notes and trouble shooting for experiments in Section B, 99
- B.II   Electrophoretic separation of DNA fragments in agarose gels, 101
- B.III   Recovery of DNA from agarose gels, 103

Section C   **DNA-dependent gene expression systems**

Introduction, 107

- C.1   λ infection of ultraviolet-irradiated cells, 109
- C.2   Maxi-cells, 114
- C.3   Mini-cells, 118
- C.4   Polypeptides synthesized from DNA templates *in vitro*, 122

Appendixes

- C.I   Some general media used in Section C, 126
- C.II   SDS-polyacrylamide gel electrophoresis, 128
- C.III   Autoradiography and fluorography of gels, 134
- C.IV   Preparation of the components of an *E. coli in vitro* transcription/translation system, 137

Section D   **Bacteriophage λ as a vector**

Introduction, 145

- D.1   Preparation of a bank of hybrid DNA, 155
- D.2   Preparation of extracts for *in vitro* packaging, 161

Appendixes

- D.I   Preparation of a gene-bank in a λ vector, 165
- D.II   DNA extraction from mouse livers, 166
- D.III   Isolation of high molecular weight bacterial DNA, 169

Section E **Electronmicroscopy of nucleic acids**

Introduction, 173

E.1 Mounting double-stranded DNA using the aqueous droplet technique, 175
E.2 Heteroduplex mapping, 184
E.3 Mapping self-annealing molecules (snap-backs), 187
E.4 Mapping by R-looping, 188
E.5 Preparation of open circles from supercoiled DNA, 190

Index, 191

Contributors
All from the University of Leicester

J. Almond BSc PhD *Department of Microbiology*
L. Beecroft BSc PhD *ICI Joint Laboratory*
E. Blair BSc PhD *Department of Biochemistry*
W.J. Brammar BSc PhD *Department of Biochemistry*
G. Baulnois BSc PhD *Department of Genetics*
D. Burt BSc PhD *ICI Joint Laboratory*
B. Holland BSc PhD *Department of Genetics*
A. Jeffreys BA DPhil *Department of Genetics*
P. McTurk BSc MPhil *Department of Biochemistry*
J. Mathews BSc PhD *Department of Botany*
P. Meacock BSc PhD *The Biocentre*
J. Pratt BSc PhD *Department of Genetics*
N. Stoker BSc PhD *Department of Genetics*
P. Williams BSc PhD *Department of Genetics*
J. Windass BSc PhD *ICI Joint Laboratory*
J. Varley BSc PhD *Department of Zoology*

Preface

This manual was written for a practical course at the University of Leicester designed to teach cloning techniques to those who had no first hand experience of them. The popularity of the course, and demand for the course manual, prompted us to edit it for publication to a wider audience.

The techniques set out in protocol form in the Manual have in common that they are all regularly used in the laboratories of the School of Biological Sciences. First hand experience of them rather than an attempt to be comprehensive was an important criterion for inclusion. Consequently the Manual does not lay claim to offer a complete coverage of the subject. A knowledge of microbiological technique and basic microbial genetic skills is assumed and a popular guide is available—*Experiments in Molecular Genetics,* by J.H. Miller, published by Cold Spring Harbor Laboratory, N.Y., 1972. Other important techniques, such as DNA sequencing were included in the course at Leicester as demonstrations backed up by lectures and seminars. No practical work was performed however and none have been added to the published manual. Recent practical reviews can be consulted by those who need these methods (Maxam & Gilbert 1980; Bankier & Borrell, 1983).

The different sections were designed and written by different groups. Consequently there were inevitably differences in style and layout, and some repetition of descriptions of commonly used techniques. In preparing the manual for publication we have endeavoured to bring a measure of uniformity to the style and layout. But we have tried to do this with a light touch so that the spontaneity of the original versions has not been edited out. Procedures do not always succeed when they are translocated to different laboratories. The introductory section which precedes each of the experiments contains references which, wherever possible, provide background information that should help to identify any problems. The authors of the various sections may also be able to help and would welcome being told of difficulties that have been encountered.

The experimental procedures are set out in protocol form with an indication of the time required to complete them. They can all be fitted into an intensive 2-week course by running different sections of a complete procedure in parallel. This was the format of the course at Leicester. Under a more leisurely time-table the longer procedures can be carried through from start to finish in sequence.

There are numerous references to specific manufacturers in the text. In many cases there will be acceptable alternative manufacturers who can supply similar products and the one cited here is simply a matter of local preference at the time of writing, or habit.

The number of strains of bacteria, and their viruses and plasmids, employed in the course is not large. Many of them are widely used and will be found in laboratories all over the world. The authors of the different sections of the course have also agreed to supply those used in their Sections to readers of this manual who cannot obtain them locally.

The list of people who have helped to teach this course on two occasions inevitably omits to name many people at Leicester who have contributed to its success. We would like to record our gratitude and thanks to them.

We are most grateful to the Biotechnology Directorate of the Science and Engineering Research Council for providing a substantial grant towards the cost of this course since 1981, and to many industrial companies who have also generously supported us. We are also greatly indebted to the many people who have helped with the development, over several years, of many difficult techniques which we are now able to present as relatively routine.

I.B. Holland
R.H. Pritchard

References

Maxam, A.M. & Gilbert, W. (1980) Sequencing end-labelled DNA with base specific chemical cleavages. *Meth. Enzym.*, **65,** 499.

Bankier, A.T. & Borrell, B.G. (1983) Shotgun DNA sequencing. *Techniques in the Life Sciences,* B5. Ch. B508; Nucleic Acid Biochem. pp. 1–34. Elsevier, Amsterdam.

Section A   Mammalian cDNA cloning

Jeff Almond
Linda Beecroft
Eric Blair
Peter Meacock
John Windass

Introduction

Section A is devoted to the synthesis and cloning of DNA copies of eukaryotic messenger RNAs (mRNAs). For an explanation of why cloning copy DNA (cDNA) is distinct from cloning genomic DNA and why it is often advantageous, see the excellent review by Williams (1981). The essential points are that mRNAs represent the contiguous protein-coding functional domain of a gene and that their relative abundance within a given cell is determined by the state of differentiation of that cell. Most eukaryotic mRNAs have a 3' polyA tract, which can be used to advantage in their purification and the subsequent synthesis of cDNA.

This section of the course attempts to cover all the essential stages in cloning cDNAs of eukaryotic mRNAs in a plasmid vector. The experiments to be carried out would normally form a continuum, and realistically could be expected to take 2–4 weeks. In order to compress them into a 4-day timetable this section has been broken into three subsections which can run concurrently. They cover the following topics:

1 Isolation, purification and characterization of polyA-containing messenger RNA from animal tissue.

Table A.1.

	Subsection A.1	Subsection A.2	Subsection A.3
Day 1			
a.m.	—	First-strand cDNA synthesis	Competent cell preparation
p.m.	RNA extraction	G50 column	Transformation
Day 2			
a.m.	(i) RNA extraction (ii) Measure absorbance	Second-strand cDNA synthesis	Colony picking
p.m.	Oligo dT-cellulose chromatography	G50 column	Colony picking
Day 3			
a.m.	mRNA precipitation	S1 nuclease treatment	Prepare filters—cell lysis
p.m.	*In vitro* translation of mRNA	(i) Homopolymer tailing—pilot (ii) cDNA gel	(i) Hybridization of filters (ii) Inoculate minicultures
Day 4			
a.m.	—	(i) Homopolymer tailing (ii) Develop cDNA autoradiograph	(i) Plasmid minipreps (ii) Wash and expose filters
p.m.	—	Annealing	(i) Digestion and gel analysis of plasmid DNAs (ii) Develop filter autoradiograph

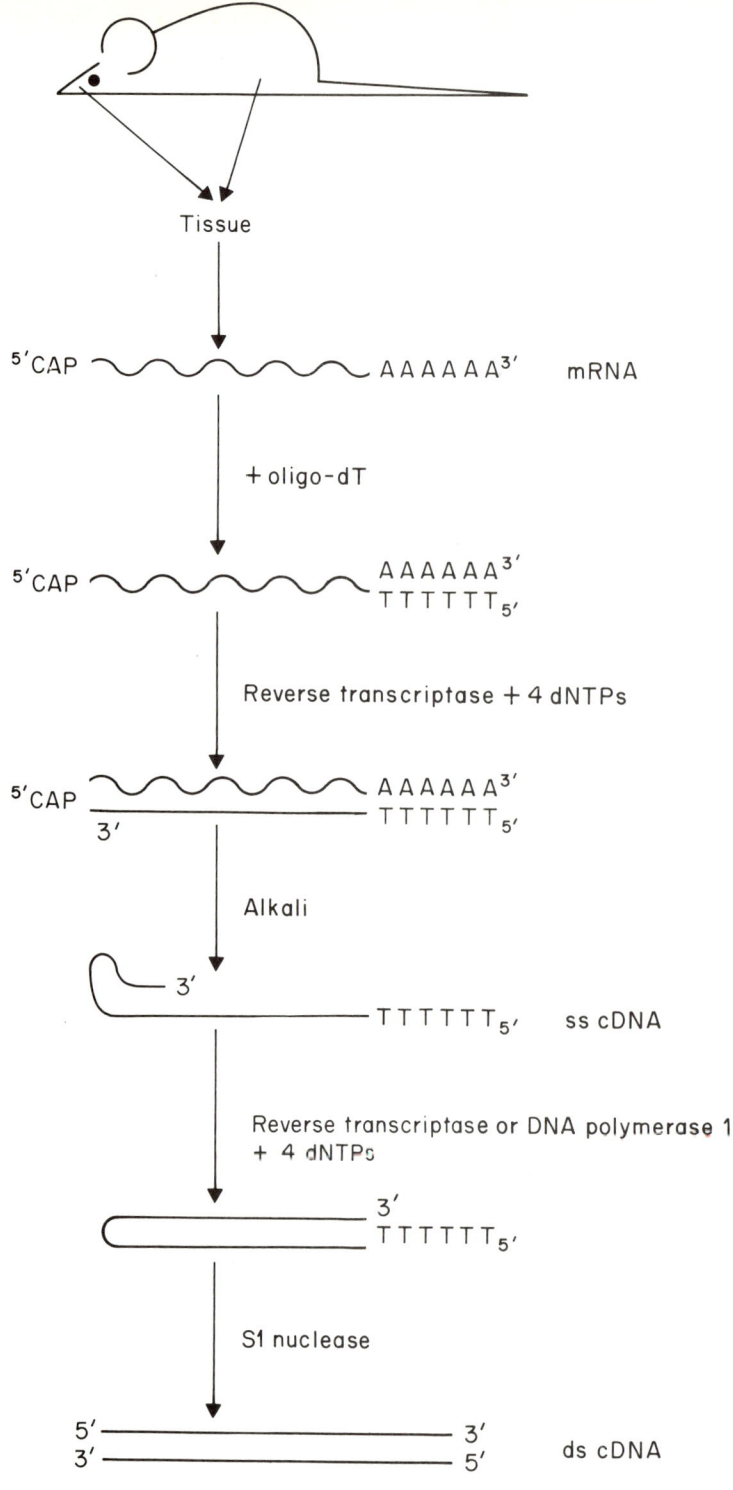

Fig. A.1 Preparation of cDNA

2 Synthesis of homopolymer-tailed double-stranded cDNA and construction of hybrid plasmids by annealing to complementary-tailed vector DNA.

3 Cloning of hybrid plasmids and analysis of clones by insertional inactivation of antibiotic-resistance genes, colony hybridization and restriction endonuclease digestion.

These stages are outlined in Figs A.1 and A.6 and a sample timetable for a 4-day programme is given in Table A.1.

Reference

Williams, J.G. (1981). The preparation and screening of a cDNA clone bank. In: *Genetic Engineering. Vol. 1.* (Ed. Williamson, R.), pp. 1–61. Academic Press, London.

Section A.1 Isolation of mRNA and translation *in vitro*

Background

Many methods are currently in use for isolation of ribonucleic acids from mammalian cells, tissues and organs. It is particularly important to obtain a good quality, undergraded mRNA preparation, since this ultimately determines the quality of cDNA synthesized and the size and usefulness of clones.

The method described here (Noyes *et al.* 1979) is particularly appropriate for isolation of mRNA from animal tissues.

To prevent degradation of the RNA by cellular ribonucleases, it is important to use either extremely fresh tissue or tissue placed immediately into liquid nitrogen at the time of collection and stored at $-70°C$ until required.

It is equally imperative to introduce the tissue to the phenol/chloroform/isoamyl alcohol (PCI) mixture as soon as possible, since this should protect the RNA from further degradation by RNases. Several PCI extractions and back extractions are required in order to achieve thorough deproteinization of the total cellular RNA. The RNA is then separated from DNA by passage through a caesium chloride 'pad' gradient. We find this method excellent for extracting RNA from soft animal tissues. However, alternative methods may be better for RNA extraction from plant cells, solid tissue or tissue culture cells. A general review of extraction methods for RNA is given by Taylor (1979).

References

Noyes, B.E., Mevarech, M., Stein, R. & Agarwal, K.L. (1979) Detection and partial sequence analysis of gastrin mRNA by using an oligodeoxynucleotide probe. *Proc. Natl. Acad. Sci. U.S.A.*, **76**, 1770.

Taylor, J.M. (1979) The isolation of eukaryotic messenger RNA. *Ann. Rev. Biochem.*, **48**, 681.

Important note

RNA is very susceptible to degradation by nucleases which can easily contaminate apparatus and solutions. Virtually all biological cells and tissues, especially human skin, are rich in these nucleases. When working with RNA the user should take every precaution to avoid the presence of nucleases. Gloves should be worn at all times and all glass apparatus used should be autoclaved sterile. It is also advisable that all water to be used for solutions containing RNA should be treated with diethyl pyrocarbonate (DEPC) before use (see p. 8).

Experiment A.1 Isolation of RNA

Materials needed (all sterile)

TLE buffer (0·2 M Tris-HCl, pH 7·5; 0·1 M LiCl; 25 mM EDTA; 0·1% SDS)
PCI mixture (50: 48: 2, phenol: chloroform: isoamyl alcohol)
CE solution (5·7 M $CsCl_2$, 0·1 M EDTA, pH 7·5)

All solutions are prepared with DEPC-treated H_2O which is produced as follows:
To 1 litre distilled water add 1 ml diethyl pyrocarbonate (DEPC). Shake well and allow to stand for 10 min.

Autoclave at 15 psi for 15–20 min to remove unreacted DEPC. This step is particularly important as any remaining DEPC will poison subsequent enzyme reactions.

Procedure

1 Mix together equal volumes of TLE buffer and PCI mixture in the homogenizer flask. (We suggest 30 ml of each for a maximum of 10 g of tissue.)
2 Weigh tissue provided as quickly as possible and add to the flask.
3 Homogenize tissue thoroughly in a Dounce (or equivalent) homogeniser.
4 Divide homogenate equally between glass universal centrifuge bottles. Shake vigorously for a few minutes. Spin at 5000 rpm for 15 min in a bench centrifuge.
5 Remove aqueous supernatant to conical flask containing 20 ml of PCI mixture and shake vigorously for 5 min.
6 Re-extract the lower phenol phase twice with 15 ml TLE buffer, each time spinning as in step 4 and pool the supernatant with that of step 5.
7 Shake the flask with the pooled supernatants vigorously for 5 min. Spin at 5000 rpm for 15 min.
8 Remove aqueous phase to sterile measuring cylinder. N.B. If the aqueous phase is not translucent at this stage carry out a further PCI extraction.
9 Carefully layer the aqueous phase over 0.2 vol CE solution. Top up the tubes with liquid paraffin if necessary, and balance the tubes. Load into a swing out rotor and centrifuge for 18 h at 25000 rpm (83 000 g) at 25°C.

10 Next day, remove paraffin and supernatant with Pasteur pipette. Identify the DNA band above the CsCl pad and remove this carefully.

11 Invert the tube, allow any remaining CsCl solution to drain away.

12 Take up the translucent RNA pellet in 1–2 ml of DEPC-treated water and allow to redissolve over about an hour. Transfer to Corex tube.

13 Add 0·1 vol 2 M sodium acetate and 2 vol absolute ethanol. Precipitate in a dry-ice bath for 30 min.

14 Centrifuge at 10 000 rpm in Sorval SS34 rotor (or equivalent) for 30 min at 0°C and carefully decant the ethanol supernatant.

15 Cover tube with Parafilm. Prick several holes and dry under vacuum for 5 min (do not over-dry). This procedure prevents loss of dry pellet through air turbulence.

16 Resuspend pellet in 0·4 ml DEPC water.

17 Remove a small aliquot (say 2·5 μl) and dilute to 500 μl. Estimate the RNA concentration by reading the absorbance at A_{260} (1 mg ml^{-1} RNA = 20 A_{260} units).

Experiment A.2 Purification of polyadenylated mRNA by affinity chromatography on oligo-dT cellulose

Background

Most eukaryotic mRNAs possess a tract of adenylic acid residues up to 200 in length at their 3' terminus (Aviv & Leder 1972). This property can be turned to use in a one-step purification of mRNA away from ribosomal RNA (rRNA) and transfer RNA (tRNA). Under high salt conditions (400 mM NaCl) the polyA tract of mRNA will hybridize to a tract of thymidylic acid residues (usually around 20 in length) covalently linked to cellulose. Ribosomal and tRNA do not possess polyA and will not bind. The mRNA can be released from hybrids by subsequently lowering the salt concentration to zero (Faust et al. 1973). Practically, some ribosomal RNA is usually found to bind to the oligo dT-cellulose column at 400 mM NaCl. This RNA fraction can be largely released before mRNA elution by inclusion of an intermediate wash step of 100 mM salt. However, the presence of small quantities of rRNA seems neither to affect *in vitro* translation nor cDNA synthesis.

References

Aviv, H. & Leder, P. (1972) Purification of biologically active globin mRNA by chromatography on oligothymidilic acid-cellulose. *Proc. Natl. Acad. Sci. U.S.A.*, **69**, 1408.

Faust, C.H., Digglemann, H. & Mach, B. (1973) Isolation of poly(adenylic-acid) rich ribonucleics acid from mouse myeloma and synthesis of complementary deoxyribonucleic acid. *Biochemistry*, **12**, 925.

Materials needed

Oligo dT-cellulose packed in a column equilibrated in binding buffer

Binding buffer (0·4 M NaCl; 10 mM Tris-HCl, pH 7·6; 0·02% SDS)

Elution buffer (1 mM EDTA; 10 mM Tris-HCl, pH 7·6; 0·02% SDS)

Acid washed quartz cuvettes

Siliconized* 15 ml Corex tubes

2× binding buffer

0·1 M NaOH

*Note: Rinse *dry* glass tubes with dichlorodimethyl silane (1%) in toluene. Wash ×5 in distilled water. Bake at 160°C for 3 h. This makes glass water-repellant.

Procedure

1   Adjust ionic strength of total RNA to that of binding buffer by adding an equal volume of $2\times$ binding buffer.

2   Apply to column and collect the eluate in a Corex tube. Re-apply this to the column. Repeat if desired.

3   Run 5 ml binding buffer through the column and collect in the Corex tube. This is the polyA$^-$ fraction and should be retained. Precipitate with 2·5 vol ethanol. Leave overnight at $-20°C$ or 1 h at $-70°C$.

4   Wash the column with a further 20 ml of binding buffer.

5   Run 5 ml of elution buffer through the column, collecting it into a fresh Corex tube. This is the polyA$^+$ fraction. Precipitate with 0·1 vol of 2 M sodium acetate and 2·5 vol ethanol. Leave overnight at $-20°C$ or 1 h at $-70°C$.

6   To regenerate the column, wash with 20 ml of elution buffer. Then run in 10 ml of 0·1 M NaOH and allow to stand for 30 min. Finally wash with binding buffer until the pH has returned to neutral.

7   Spin Corex tubes at 11 000 rpm for 30 min at 0°C and carefully decant ethanol supernatants. Vacuum dry as previously described (see p. 9).

8   Resuspend polyA$^-$ fraction in 0·5 ml and polyA$^+$ fraction in 250 μl. Estimate the respective RNA concentration by reading the A_{260} of small aliquots diluted to 0·5 ml. Alter concentration to 1 mg/ml if required.

9   If time allows pass half the polyA$^+$ fraction through a second oligo dT column (i.e. repeat steps 1–5). This yields the polyA^{++} mRNA which stimulates *in vitro* translation far more efficiently than polyA$^+$, and is also a far better substrate for cDNA synthesis.

Experiment A.3 *In vitro* translation of mRNA

Background

In vitro translation provides a powerful method for assaying specific mRNAs, for monitoring their purification, for studying their properties, and for identifying sequences complementary to mRNA in cDNA plasmid clones. All these approaches are essentially variations on a basic theme which is illustrated here using the rabbit reticulocyte protein-synthesizing system and mRNA.

Many considerations enter into the choice of cell-free system and experimental approach. Broadly, the two types of cell-free system favoured by most investigators are the reticulocyte system, demonstrated here, and a cell-free system prepared from wheat germ (Roberts & Patterson 1973). Messenger RNAs which can be well translated in one system may be poorly translated in the other, for no readily apparent reason. The reticulocyte system may be better for synthesis of large proteins (up to 200 000 daltons) than wheat germ; wheat germ may possess more endogenous ribonucleases than reticulocytes.

The products of translation can be analysed readily by gel electrophoresis and autoradiography. Possession of a specific antibody raised against the protein under study can be used in an immunoprecipitation assay of translation products (Parnes *et al.* 1981). Here, we assess translation efficiency by incorporation of radioactive counts into TCA precipitable material.

Preparation of the reticulocyte system is briefly described in Appendix A.I

References

Parnes, J.R., Velan, B., Felsenfeld, A., Ramanathan, L., Ferrini, U., Appella, E. & Sudman, T.G. (1981) Mouse B2-microglobulin cDNA clones: A screening procedure for cDNA clones corresponding to rare mRNAs. *Proc. Natl. Acad. Sci. U.S.A.*, **78**, 2253.

Roberts, B.E. & Paterson, B. (1973) Efficient translation of Tobacco Mosaic Virus RNA and rabbit globin 9S RNA in a cell-free system from commercial wheatgerm. *Proc. Natl. Acad. Sci. U.S.A.*, **70**, 2330.

Materials needed

Rabbit reticulocyte lysate (commercially available or prepared as in Appendix A.I) micrococcal nuclease treated and stored at $-70°C$.

^{35}S-methionine (specific activity > 1,000 Ci mmol^{-1})
Messenger RNAs (approx. 1 mg ml^{-1} in H$_2$O), stored $-70°C$
Glass capillary micropipettes, sterile, 1–5 µl.

Procedure

1 Quickly thaw out lysate, mRNAs, and ^{35}S-methionine just before addition by placing in 37°C water bath. Quickly re-freeze all solutions after use in ethanol/dry ice and return to appropriate freezer. (Use aliquots of reticulocyte lysate only once—do not re-freeze and re-use since translational activity is lost upon repeated freezing and thawing.)

2 Mix the following in an Eppendorf tube sequentially, initiating reaction by additions of mRNA and ^{35}S-methionine: Lysate, 10µl; ^{35}S-methionine, 2 µl; mRNA, 0–1 µl; H$_2$O to 13 µl. Work quickly.

Set up six reaction mixtures with:
  no mRNA.
  0.5 µl quality control RNA (usually commercially provided).
  0.5 µl and 1.0 µl aliquots of polyA$^+$ and polyA^{++} if prepared.
  0.5 µl and 1.0 µl of polyA$^-$RNA.

3 Vortex the tubes and quickly spin in Eppendorf microfuge (2 s).

4 Incubate tubes at 30°C for 60 min.

5 Add a 2 µl aliquot of retriculocyte cell-free translation products to 0.5 ml 1 M NaOH in 10% H$_2$O$_2$. Mix and incubate at 37°C for 30 min. (H$_2$O$_2$ discolours the haem which causes quenching in the scintillation spectrometer and the NaOH hydrolyses charged amino acyl tRNA.)

6 Add 1.0 ml 20% trichloracetic acid.

7 Leave on ice for 15 min. Soak Whatman 2.5 cm GF/C filters in 10% TCA.

8 Filter samples on filtration manifold, washing with 3 × 5 ml 10% TCA, 1 × 5 ml ethanol.

9 Dry. Count.

The polypeptide products of the reaction just completed will also be analysed by SDS-PAGE. This technique is described in detail in Section C.

Section A.2 Synthesis of double-strand cDNA and the construction of hybrid pAT153/cDNA plasmids by the homopolymeric tailing method

Introduction

The objectives of this series of experiments are:
1 The synthesis of single-stranded cDNA.
2 The conversion of single-stranded cDNA into double-stranded cDNA.
3 The removal of any single-stranded regions of the double-stranded product.
4 The addition of complementary 3'-homopolymer tails to double-stranded cDNA and linearized pAT153 vector plasmid DNA.
5 The annealing of tailed plasmid vector and cDNA to generate recombinant plasmids which may be recovered by transformation.

The strategy is shown in outline in Figs A.1 and A.6.

Experiment A.4 First-strand cDNA synthesis

Background

Complementary cDNA is most conveniently prepared from eukaryotic mRNA preparations using avian myeloblastosis virus (AMV) reverse transcriptase (RNA-dependent DNA polymerase). The reaction conditions chosen for reverse transcription have critical effects on the nature of the products (Retzel et al. 1980). For example, low deoxynucleoside triphosphate levels, although valuable for synthesis of high specific activity cDNA probes for hybridization analyses, tend to result in short cDNA molecules unsuitable for use in molecular cloning. High salt concentrations are optimal for enzyme activity but may stabilize secondary structures in mRNA molecules which may result in incomplete reverse transcription. Actinomycin D can be used to ensure that no second-strand synthesis occurs, whilst specific oligonucleotides may be used for synthesis of particular cDNA species from mixed mRNA populations (Chan et al. 1979).

It may be advisable to optimize the conditions for specific mRNA preparations. Those given here have been found to be generally satisfactory for the synthesis of high molecular weight cDNA; they closely resemble many published procedures.

In this practical two reactions will be performed in parallel: a large scale synthesis of low specific activity ^{3}H-labelled cDNA which can subsequently be used for second strand synthesis and cloning and a small scale synthesis of α-^{32}P-dCTP labelled cDNA for electrophoretic characterization of the single-stranded reverse transcription products of your mRNA. See Appendix A.IV.

References

Chan, S.J., Noyes, B.E., Agarwal, K.L. & Sheiner, D.F. (1979) Construction and selection of recombinant plasmids containing full-length complementary DNAs corresponding to rat insulins I & II. *Proc. Natl. Acad. Sci. U.S.A.*, **76**, 5036.

Retzel, E.F., Collet, M.S. & Faran, A.T. (1980) Enzymatic synthesis of deoxyribonucleic acid by the avian retrovirus reverse transcriptase *in vivo*; optimum conditions required for transcription of large ribonucleic acid templates. *Biochemistry*, **19**, 513.

Material needed

5× RT-1 buffer (250 mM Tris-HCl, pH 8·3; 250 mM NaCl; 40 mM $MgCl_2$; 5 mM dATP; 5 mM dGTP; 5 mM TTP; 5 mM dCTP)
TE buffer (10 mM Tris-HCl, pH 7·5; 1 mM EDTA)
0·1 M dithiothreitol (DDT)
0·4 M EDTA, pH 8·0
10% sodium dodecyl sulphate (SDS)
Redistilled phenol (saturated with 0·5 M Tris-HCl, pH 7·5)
1 mg ml^{-1} oligo (dT)$_{12-18}$ in H_2O
10 μCi lyophilized ^{3}H-dCTP (15–30 Ci nmol^{-1})
5 μCi α-^{32}P-dCTP (2,000–3,000 Ci nmol^{-1})
Reverse transcriptase*
1 mg ml^{-1} polyA$^+$ mRNA in H_2O
10 mg ml^{-1} yeast tRNA in H_2O
Sterile distilled H_2O
Siliconized glass pipettes
Sephadex G50 Fine—preswollen in TE buffer
Scintillation cocktail suitable for aqueous samples
1 M NaOH
1 M Tris, pH 7·5
2 M acetic acid.

(a) Large scale single-stranded cDNA synthesis

Procedure

1 Carefully dissolve ^{3}H-dCTP in 20 μl 5×RT-1 buffer.
2 Add 10 μl 0·1 M DTT, 10 μl 1 mg ml^{-1} oligo (dT) and 5–10 μg mRNA.
3 Make up to 100 μl with sterile distilled water.
4 Add ~50 units reverse transcriptase
5 Mix carefully by gentle pipetting.
6 Incubate 60 min at 42°C. (After this incubation, dilution with an equal volume of 5 mM Tris −HCl (pH 8·3), followed by addition of a further 50 units reverse transcriptase and incubation for a further 30 min at 47°C, may allow more complete reverse transcription of certain mRNA species (e.g. those with a particularly stable secondary structure).
7 Stop reaction by addition of 2 μl 0·4 M EDTA, 2 μl 10% SDS and 100 μl Tris-saturated phenol.
8 Mix by vigorous shaking (1–2 min).

---

*Definitions of units of reverse transcriptase vary widely. One unit of enzyme obtained from Life Sciences Inc. is defined as incorporating 1 nmol dTMP into an acid insoluble product in 10 min at 37°C using a polyrA/dT$_{12-18}$template/primer. Preparations of reverse transcriptase can be obtained from Life Sciences Inc., 1509½ 49th Street, St Petersburg, Florida 33707, U.S.A.

9 Centrifuge in a bench-top centrifuge for 1 min at 15 000 g.
10 Carefully remove and retain aqueous (upper) phase.
11 Add 50 μl TE buffer to phenol phase. Mix and re-centrifuge.
12 Pool aqueous phases.

(b) Small scale single-stranded cDNA synthesis

Procedure

1 To 5 μCi α-^{32}P-dCTP add 2 μl 5×RT-1 buffer, 1μl 0·1 M DTT, 1 μl 1 mg ml^{-1} oligo dT and 0·5 μg mRNA. Make up to 10μl with sterile distilled water. Add 5 units reverse transcriptase.
3 Incubate 60 min at 42°C.
4 Stop reaction by addition of 0·5 μl 0·4 M EDTA, 0·5 μl 10% SDS, 1 μl 10 mg/ml yeast tRNA (carrier), 40μl TE and 50 μl phenol.
5 Mix, centrifuge, remove aqueous phase, re-extract phenol phase with TE and pool aqueous phases, as for large scale cDNA preparation.

(c) Separation of single-stranded cDNA from unincorporated nucleotides

Separation of large scale cDNA preparations from unincorporated nucleotides and other components of the first strand synthesis and extraction procedures is conveniently achieved on a Pasteur pipette—Sephadex column (Fig. A.2).

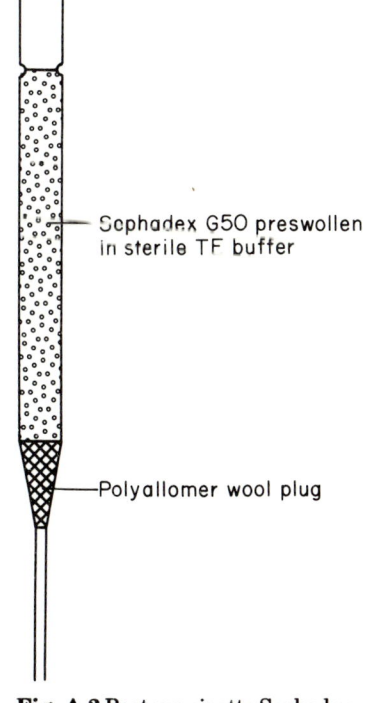

Fig. A.2 Pasteur pipette Sephadex column

Procedure

1 Take siliconized Pasteur pipettes, plug with a small wad of siliconized glass wool or polyallomer wool pushed into place with a long Pasteur pipette. Fix pipettes upright (e.g. hold with rubber bands on a ruler in a clamp stand). Fill pipettes to the constriction with Sephadex G50 Fine (preswollen in TE buffer). When the liquid head on the column reaches the packed Sephadex volume, flow will stop. Ensure the samples are clear (warm gently if necessary) and apply to the columns. Start collecting into 1·5 ml conical polypropylene tubes immediately. Wash sample in with 100 μl TE, allow column to drain. Change collection tubes. Apply 250 μl TE to the column. Allow to drain. Repeat until fifteen 250 μl fractions have been collected.

To monitor the elution profile from columns on which ^{3}H-labelled material has been fractionated, mix 5 μl aliquots from fractions 1–15 with a suitable aqueous scintillation cocktail and count in the ^{3}H channel of a scintillation counter. To monitor ^{32}P-labelled material, place tubes in scintillation vials, cap and count by Cerenkov radiation in the ^{3}H channel of a scintillation counter. The cDNA peaks will usually be in fractions 4–6. Pool appropriate fractions. Use counts eluting from columns to estimate cDNA yields and hence reverse transcription efficiencies (see example calculation Table A.2).

Note: DNA occasionally sticks to G50 columns; if this happens, washing the column with 0.5 M NaCl can sometimes recover the DNA. However, if it has not been eluted after washing with 4 ml it is probably irretrievable.

Table A.2 Specimen calculation (Large scale first-strand cDNA synthesis)

Fraction number	Typical counts (cpm)
1	13
2	14
3	18
4*	2,614
5*	5,036
6*	1,312
7	23,931
8	146,579
9	259,006
10	187,614
11	89,639
12	18,170
13	2,604
14	315
15	117
Total	736,982

Final concentration cold dCTP = 1 mM
Reaction volume = 100 μl
Therefore 100 nmol dCTP present

Total dpm recovered from column (x) = 736,982
Therefore dpm recovered/nmol dCTP (z) = 7,370

Counts incorporated into cDNA peak (y) = 4*+5*+6* = 8,962 dpm
Therefore

$= \dfrac{y}{z} = \dfrac{8,962}{7,570} = 1.216$ nmol

Since 1 nmol dCTP ≡ 0.307 μg and assuming that the four bases are equally represented in cDNA, then weight of cDNA produced (in μg)

$\approx 1.216 \times 0.307 \times 4 = 1.49$ μg

If started with 10 μg mRNA

Percentage yield
$= \dfrac{1.49}{10} \times 100 = 14.9\%$

(d) Alkali treatment of single-stranded cDNA

Before the second-strand cDNA synthesis, it is important to remove residual RNA, which might still be hybridized to the single-stranded cDNA or which might be copied in the second-strand reaction, using the first-strand cDNA as a primer. This is achieved at this stage by alkaline hydrolysis.

Procedure

1 Add 0.25 vol (x μl) of 1M NaOH to single-stranded cDNA preparations in order to give a final concentration of 0.2 M NaOH.
2 Incubate 70°C for 15 min.
3 Add x μl 1 M Tris, pH 7.5. Mix.
4 Add x μl 2 M acetic acid. Mix.
5 Add 2.5 vol absolute ethanol. Dispense into 1.5 ml conical polypropylene tubes and place at −20°C overnight or in a solid CO_2/ethanol bath for 15 min.

(e) Recovery of ethanol precipitates of cDNAs

Procedure

Copy DNA preparations can be recovered by centrifugation in a bench-top microcentrifuge in a cold room for 15 min at 15 000 g.

Carefully decant supernatants. Dry pellets for 10–15 min under vacuum (cover tubs with Parafilm and punch holes with a syringe needle before placing under vacuum).

Re-dissolve ^{32}P-labelled cDNA pellet from the small scale preparation in 10 µl H$_2$O, store frozen until analysed electrophoretically (see Appendix A.IV). Re-dissolve ^{3}H-labelled cDNA pellet in 50 µl H$_2$O. Mix 5 µl aliquot with a suitable scintillation cocktail, count to confirm efficiency of ethanol precipitation and check yield of single-stranded ^{3}H-labelled cDNA.

Experiment A.5 Second-strand cDNA synthesis

Background

An even greater variety of methods are commonly used for this reaction than for first-strand synthesis. For example, it is possible to use either *E. coli* DNA polymerase I (either Kornberg enzyme or Klenow fraction—Wickens *et al.* 1978) or the DNA-dependent DNA polymerase activity of AMV reverse transcriptase (Ullrich *et al.* 1977). Specific oligonucleotide primers may be used, if appropriate details of the sequence of the 5' terminus of the mRNA being copied are known (Houghton *et al.* 1980) or if a suitable homopolymeric tail has been added to the 3' end of the single-stranded cDNA (Land *et al.* 1981). Alternatively, the ability of single-stranded cDNA molecules to act as 'self primers' by hairpin loop formation, through localised regions of homology between the extreme 3' end of the cDNA and regions close to it, can be exploited (Rougeon & Mach 1976). Reaction conditions again vary widely but the principles of high oligonucleotide concentrations and high temperatures for maximum length transcripts still hold (for detailed review see Williams 1981).

The method described here is dependent on self-priming under conditions described by Kay *et al.* (1980), which have been found to give very satisfactory results. Again, it is suggested that two reactions be performed in parallel to give a large amount of low specific activity ^{32}P-labelled double-stranded cDNA, suitable for cloning, and a smaller amount of higher specific activity ^{32}P-labelled material, for electrophoretic analysis (see Appendix A.IV). Apart from the specific activity of ^{32}P-dCTP used in the two reactions, reaction conditions are identical.

Note. An alternative method of second-strand synthesis, using DNA polymerase 1 is outlined in Appendix A.II.

References

Houghton, M., Stewart, A.G., Doel, S.M., Emtage, J.S., Eaton, M.A.W., Smith, J.C., Patel, T.P., Lewis, H.M., Porter, A.G., Birch, J.R., Cartwright, T. & Carey, N.H. (1980) The amino-terminal sequence of human fibroblast interferon as determined from reverse transcripts obtained using synthetic oligonnleotide primers. *Nucl. Acid. Res.,* **8,** 1913.

Kay, R.M., Harris, R., Patient R.K. & Williams, J.G. (1980) Molecular cloning of cDNA sequences coding for the major α & β globin polypeptides of adult *Xenopus laevis. Nucl. Acid. Res.,* **8,** 2691.

Land, H., Grez, M., Hauser, H., Lindenmaier, W. & Schutze, G. (1981) 5' terminal sequences of eucaryotic mRNA can be cloned with high efficiency. *Nucl. Acid. Res.*, **9**, 2251.

Rougeon, F. & Mach, B. (1976) Stepwise biosynthesis in vitro of globin genes from globin mRNA by DNA polymerase of avian myeloblastosis virus. *Proc. Natl. Acad. Sci. U.S.A.*, **73**, 3418.

Wickens, M.P., Buell, G.N. & Schimke, R.T. (1978) Synthesis of double-stranded DNA complementary to lysozyme, ovomucoid and obalbumin mRNAs. *J. Biol. Chem.*, **253**, 2483.

Williams, J.G. (1981) The preparation and screening of a cDNA clone bank. In: *Genetic Engineering, Vol. 1.* (Ed. Williamson, R.), pp. 1–61. Academic Press, London.

Ullrich, A., Shine, J., Chirgwin, J., Pictet, R., Fischer, E., Rutter, W.J. & Goodman, H.M. (1977) Rat insulin genes: construction of plasmids containing the coding sequences. *Science*, **196**, 1313.

(a) Large scale second-strand cDNA synthesis

Try to use 1 µg single-stranded cDNA as a template; **reduce reaction volumes accordingly if less is available.**

Materials needed

5× RT-2 buffer (250 mM Tris-HCl, pH 8·3; 40 mM $MgCl_2$; 0·1 M DTT; 2 mM dATP; 2 mM dGTP; 2 mM TTP; 2 mM dCTP)
0·4 M EDTA, pH 8·0
10% SDS
Redistilled phenol (0·5 M Tris-HCl, pH 7·5, saturated)
49:1 chloroform/isoamyl alcohol (0·5 M Tris-HCl, pH 7·5, saturated)
TE buffer
3 M sodium acetate, pH 6·5
5 µCi and 10 µCi of α-^{32}P-dCTP (2–3,000 Ci nmole^{-1})
Reverse transcriptase

Procedure

1 Add to 5 µCi α-^{32}P-dCTP: 100 µl 5× RT-2 buffer, 1 µg single-stranded ^{3}H-labelled cDNA, H_2O to 500 µl, and 180–200 u reverse transcriptase.
2 Mix carefully by gentle pipetting.
3 Incubate 3 h at 46°C.
4 Stop reaction by addition of 10 µl 0·4 M EDTA, 10 µl 10% SDS and 500 µl phenol.
5 Mix thoroughly by vigorous shaking.
6 Centrifuge 1 min in bench-top mini-centrifuge.
7 Pipette off aqueous (upper) phase.
8 Add 500 µl $CHCl_3$/IAA, mix thoroughly, centrifuge and remove aqueous (upper) phase.

Table A.3 Specimen calculation (Large scale second-strand cDNA synthesis)

Fraction number	Typical counts (cpm)
1	40
2	48
3	72
4	178
5	5,944*
6	838
7	19,773
8	305,544
9	740,587
10	595,415
11	129,395
12	12,979
13	1,087
Supernatant	870,150
Total	2,682,050

Final concentration of cold dCTP = 0·4 mM

Reaction volume = 500 µl
 number of nmol dCTP present = 200 nmol

Total cpm recovered = 2,682,050
 cpm recovered/nmol dCTP = 13,410

Counts incorporated into cDNA = 5,944
 nmol cDNA synthesised =
 $\frac{5,944}{13,410} = 0.443$ nmol.

$\equiv 0.443 \times 0.307 \times 4 = 0.55$ µg cDNA
 Yield of double-stranded cDNA =
 $2 \times 0.55 = 1.1$ µg

*cDNA.

9 Divide into two equal aliquots in 1·5 ml conical polypropylene tubes, add 25 µl 3 M sodium acetate and 625 µl absolute ethanol to each tube.
10 Mix thoroughly.
11 Place in solid CO_2/ethanol bath for 15 min.
12 Centrifuge 15 min in cold room in bench-top mini-centrifuge.
13 Carefully decant very radioactive supernatants and retain for counting.
Care! Pellet will also be very radioactive.
14 Drain pellets carefully.
15 Redissolve pellets in TE (125 µl each), pool two solutions.
16 Fractionate sample on G50 Sephadex/Pasteur pipette column, as in single-strand synthesis.
17 Monitor 250 µl fractions by Cerenkov radiation.
18 Pool excluded double-stranded cDNA fractions, add 0·1 vol 3 M sodium acetate and 2·5 vol absolute alcohol, store overnight $-20°C$.
19 Use counts from columns to estimate yields (see example calculation Table A.3).

(b) Small scale second-strand cDNA synthesis

May be carried out concurrently with large scale synthesis.

Procedure

1 Add to 0·1 µg ^{3}H-labelled single-stranded cDNA: 10 µCi ^{32}P-dCTP, 10 µl 5× RT-2 buffer, 0·1 µg single-stranded ^{3}H-labelled cDNA, H_2O to 50 µl and 18–20 units of reverse transcriptase..
2 Mix carefully.
3 Incubate for 3 h at 46°C.
4 Stop reaction by addition of 1 µl 0·4 M EDTA, 1 µl 10% SDS and 50 µl phenol.
5 Mix thoroughly, centrifuge, remove aqueous phase.
6 Add 50 µl $CHCl_3$/IAA, mix thoroughly, centrifuge, remove aqueous phase.
7 Add 5 µl 3 M sodium acetate and 125 µl absolute ethanol, mix.
8 Place in solid CO_2/ethanol bath for 15 min.
9 Centrifuge for 15 min in cold room in bench-top mini-centrifuge.
10 Carefully remove *very radioactive* supernatant and retain for counting.
Care! Pellet will also be very radioactive.

11 Drain pellet carefully.
12 Redissolve in 100 μl TE.
13 Fractionate sample on Sephadex G50 Fine/Pasteur pipette column, as previously.
14 Monitor 250 μl eluate fractions by Cerenkov radiation. Estimate cDNA yield as indicated.
15 Pool excluded double-stranded cDNA peak, add 0·1 vol 3 M sodium acetate and 2·5 vol absolute alcohol, store overnight −20°C.

(c) Recovery of overnight ethanol precipitate of double-stranded cDNAs

This is achieved by the same method as is used for single-stranded cDNA preparations.

Procedure

1 Redissolve the higher specific activity ^{32}P-labelled cDNA in 10 μl H$_2$O.
2 Store −20°C.
3 Redissolve the large scale cDNA preparation in 100 μl H$_2$O. Take 5 μl aliquot and mix with 250 μl H$_2$O in 1·5 ml conical polypropylene centrifuge tube, Cerenkov count to confirm efficiency of ethanol precipitation and check yield of second-strand ^{32}P-labelled cDNA (the amount of double-stranded cDNA obtained may be considered to be twice the amount of second-strand synthesised).

Experiment A.6 S1 nuclease treatment of double-stranded cDNA

Background

Double-stranded cDNA preparations synthesized in the manner described in this manual are of 'hairpin' structures with a variable sized single-stranded, covalently joined loop at one end. Varying amounts of completely single-stranded cDNA or single-stranded tails on double-stranded molecules may also be present. These single-stranded regions must be removed before cloning. This is most conveniently achieved by the use of the single-strand specific S1 nuclease derived from *Aspergillus oryzae*.

S1 nuclease preparations from commercial sources vary considerably in their specificity, some lead to quite rapid exonucleolytic cleavage of double-stranded cDNA. Each new batch, therefore, must be carefully characterized before use on cDNA preparations for cloning; a suggested procedure is given in the Appendix A.III.

References

Shenk, T.E., Rhodes, C., Rigby, P.W.J. & Berg, P. (1975) Biochemical method for mapping mutational alterations in DNA with S1 nuclease: the location of deletions and temperature-sensitive mutations in simian virus 40. *Proc. Natl. Acad. Sci. U.S.A.*, **72**, 989.

Vogt, V.M. (1973) Purification and further properties of single-strand specific nuclease from *Aspergillus oryzae*. *Eur. J. Biochem.*, **33**, 192.

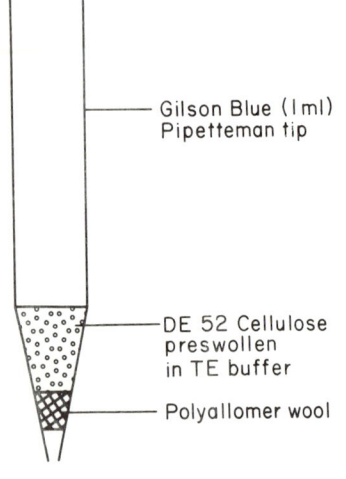

Fig. A.3 DE 52 cellulose column.

Materials needed

$5 \times$ S1 buffer (1·4 M NaCl; 22·5 mM $ZnCl_2$; 0·15 M Na acetate, pH 4·5).

SI nuclease

SDS (10%)

EDTA (0·4 M, pH 8·0)

Tris-saturated redistilled phenol.

DE 52 cellulose preswollen in TE buffer. Approximately 200 µl bed volume column poured in a Blue (1 ml) Gilson Pipetteman tip (or equivalent) plugged with polyallomer wool plug and with the bottom 2–3 mm then cut off. The column is supported over the collection tubes with a 'snap-back' clip—see diagram, Fig. A.3.

Elution buffer: 1 M NaCl; 50 mM Tris-HCl, pH 75; 1 mM EDTA.

Procedure

1 Place approximately 500 ng of the large scale double-stranded cDNA preparation in a 1·5 ml polypropylene tube on ice.
2 Add 40 μl 5×S1 buffer, up to 200 units S1 nuclease*, and H_2O to 200 μl final volume. (Reduce volume and enzyme proportionately if less cDNA is available.)
3 Incubate for 1 min at 37°C.
4 Add 5 μl 0.4 M EDTA, 5 μl 10% SDS and 200 μl Tris saturated phenol.
5 Mix. Centrifuge. Collect aqueous phase (upper layer). Re-extract phenol phase with 50μl TE and pool aqueous phases as previously.
6 Run Sephadex G50 column as previously described.

If cDNA is ethanol-precipitated directly after the Sephadex chromatography (step 6) it is sometimes a poor substrate for terminal transferase tailing. This problem can be overcome by the following procedure:

7 Pool Sephadex-excluded double-stranded cDNA fractions.
8 Apply to a small DE 52 cellulose column.
9 Allow entire Sephadex-excluded fraction to run onto the column (these columns stop flowing when the head volume reaches the bed volume of the DE 52 cellulose).
10 Wash the column with 1 ml TE.
11 Elute cDNA with 2×200 μl elution buffer into two 1·5 ml conical polypropylene tubes.
12 Add 1 ml absolute alcohol.
13 Place tubes in a solid CO_2/ethanol bath for 15 min.
14 Warm tubes slightly to reduce viscosity and centrifuge for 15 min in bench-top centrifuge in the cold room.
15 Carefully decant supernatant, drain, cover tube with Parafilm. Punch several holes in the Parafilm and dry under vacuum.
16 Dissolve cDNA pellet in 21 μl of water.
17 Take 1 μl cDNA solution and 5 ml scintillation fluid and count in ^{32}P channel of a liquid scintillation counter.
18 Calculate the yield of double-stranded cDNA in μg. Assuming a mean cDNA size of 1,000 base pairs, the yield in pmol of double-stranded cDNA and hence the number of pmol ends present can be determined (see Table A.4).

Table A.4 Calculation of the yield of double-stranded cDNA in pmol

Assume average size of double-stranded cDNA = 1,000 bp.
Molecular weight = 660,000
1 pmol = 0·66 μg
0·66 μg ≡ 2 pmol ends

*It is advisable to determine on a small aliquot of double-stranded cDNA if this amount of a particular S1 nuclease preparation can be used safely (see Appendices A.III and A.IV: Fig. A.5).

Sample preparation and electrophoresis

1 To the cDNA sample add an equal volume of formamide-dye solution.
2 Denature the sample by boiling for 2 min in a waterbath and quick chilling in an ice-water bath.
3 Flush out the sample wells of the gel and load the samples. (A Hamilton microsyringe may be useful—if so, flush out the syringe between samples with buffer.)
4 Load some ^{32}P-labelled molecular weight marker DNAs into adjacent slots.
5 Electrophorese at approximately 45 mA until the xylene cyanol dye (light blue) has migrated ½ to ⅔ down the gel.
6 Dismantle the apparatus, carefully separate the plates and cover the gel with Clingfilm (eliminating all air bubbles).
7 Autoradiograph the gel as described in Appendix C.III.

Typical examples of the results obtained are shown in Figs A.4 and A.5.

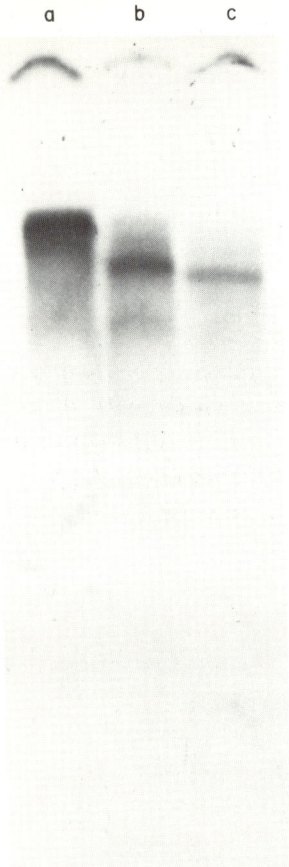

Fig. A.5 Autoradiography of S1-treated ^{32}P-labelled cDNA samples.
^{32}P-labelled cDNA samples were eletrophoresed through a 5% polyacrylamide-urea denaturing gel as described above.
Samples
Double-stranded cDNA prepared from mouse salivary gland mRNA, as in Experiments A.4 and A.5 was treated with S1 nuclease (Experiment A 6) for: (a) 0 min; (b) 1 min; (c) 5 min.

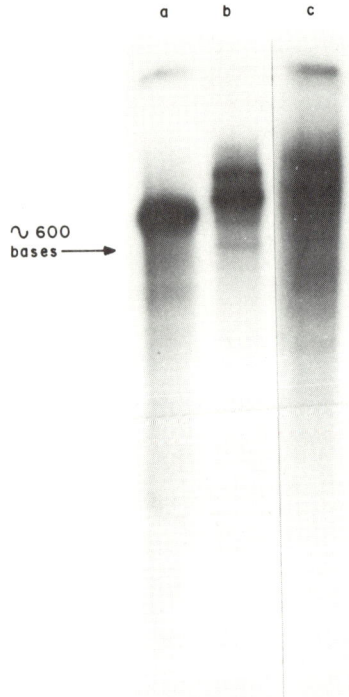

Fig. A.4 Autoradiograph of ^{32}P-labelled cDNA preparations.
^{32}P-labelled cDNA preparations were electrophoresed through a 5% polyacrylamide-urea denaturing gel as described in the text.
Samples
(a) Single-stranded cDNA synthesized as in Experiment 4 from rabbit reticulocyte 'globin' mRNA.
(b) Single-strand cDNA synthesized as in Experiment from mouse salivary gland mRNA.
(c) Double-stranded cDNA synthesized as in Experiments A.4 and A.5 from mouse salivary gland mRNA.

Experiment A.7 Homopolymeric tailing of double-stranded cDNA

Background

Having prepared double-stranded cDNA, it is usual to use one of two methods to create 3' termini which are complementary to the termini of a linearized vector. These methods are (1) addition of DNA linkers which contain restriction endonuclease cleavage sites and (2) addition of complementary homopolymer tracts to the cDNA and cloning vector (Jackson

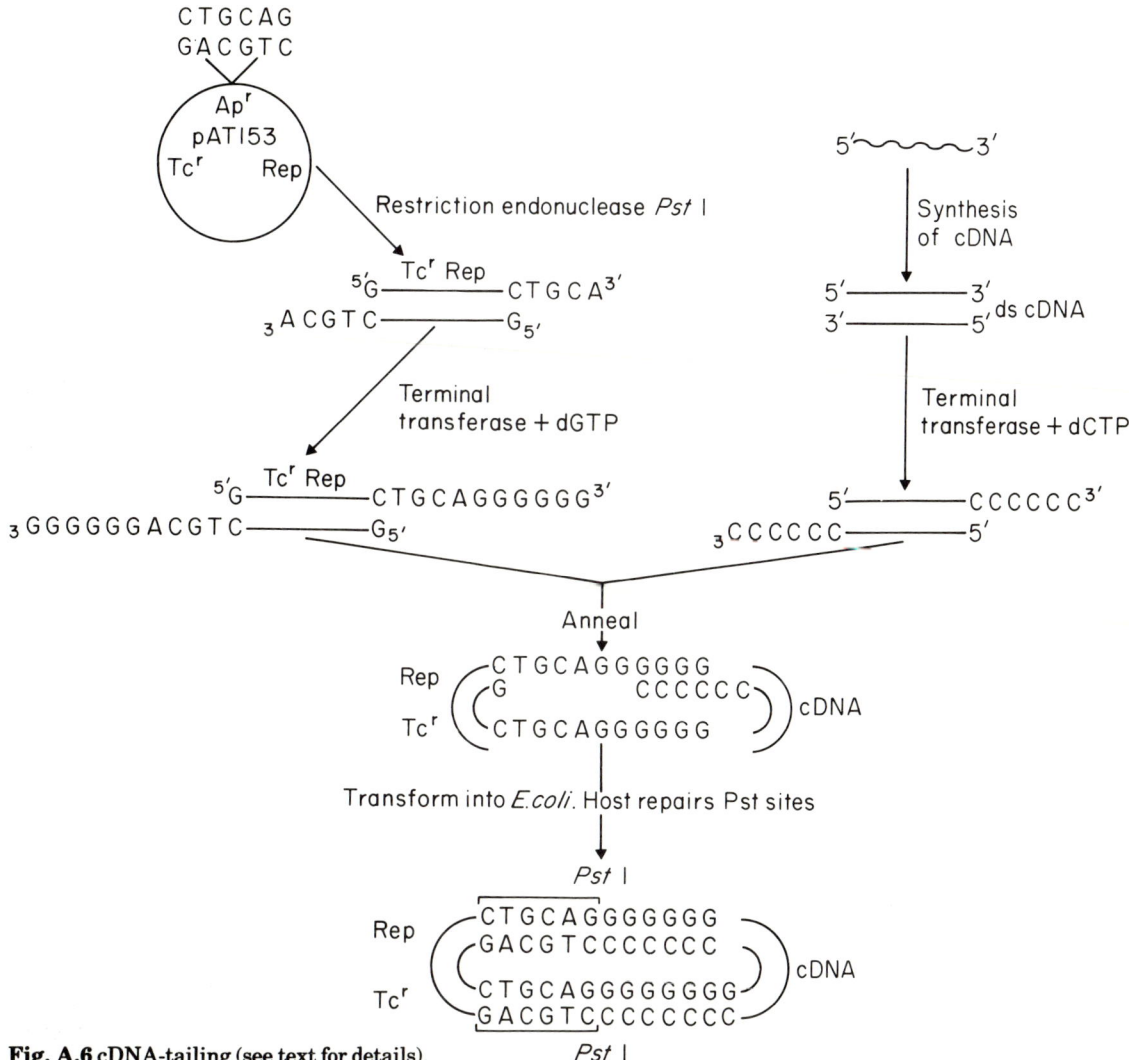

Fig. A.6 cDNA-tailing (see text for details).

et al. 1972, see also Williams 1981). Method 2 is described here since this has been most widely used because of its lower cost and independence of enzymatic ligation. Addition of poly dC tracts to cDNA synthesized *in vitro*, followed by annealing to pAT153 or pBR322 poly dG tailed after cleavage at the *Pst1* site in the ampicillin resistance gene, is presently the method of choice. This procedure has the advantage of regenerating *Pst1* cleavage sites at each end of the inserted cDNA thus enabling its precise excision (illustrated in Fig. A.6.). The interruption in the ampicillin-resistance gene renders the majority of *E. coli* clones harbouring recombinant plasmids ampicillin sensitive. This is useful for screening purposes (see below). Fig. B.1 shows a map of pAT153 and pBR322.

Addition of homopolymer tracts to 3′ termini of DNA is carried out using the enzyme terminal deoxynucleotide transferase. The nature of the 3′ terminus presented to the enzyme is important in the reaction; a protruding 3′ terminus being the best substrate and a recessed 3′ terminus the worst. The presence of cobalt in the reaction mix goes some way to alleviating this disparity and is now almost universally used (Roychoudry *et al.* 1976). The optimal length of tails on cDNA and plasmid vector is around 20. A detailed investigation of the tailing reaction is given by Deng and Wu (1982).

The experiment can be divided into two parts: (i) preparation of poly dG tailed vector DNA and (ii) poly dC tailing of cDNA.

The vector may be prepared well in advance of any cloning experiment and it is important to have strict quality controls. First, the vector DNA must be free of any contaminating *E. coli* chromosomal DNA, as a dirty batch of vector may lead to the generation of artefacts upon cloning. In practice we use pAT153 which has been through two CsCl gradient purification runs. Secondly, *Pst1* digestion of the vector must be complete. The transformation frequency of competent *E. coli* with linear vector should ideally be around $10^{-4} \times$ the frequency with supercoiled DNA and this usually requires an enzyme unit excess over the amount of DNA (e.g. 2 units/μg of DNA for 2 h). It is important to use a high quality enzyme for this purpose and not to overdigest, since trace amounts of contaminating exonucleases in restriction enzyme preparations may nibble away the 3′ protrusion and prevent subsequent regeneration of the *Pst1* recognition site. Ideally the digestion of vector should be monitored by agarose gel analysis, and by transformation, before it is used for cloning experiments. Preservation of the reconstruction site can be monitored by ligation and recutting of the digested plasmid. Enough vector can be made in a single experiment to carry out many cloning experiments. It is therefore really worthwhile to be particularly fastidious with this procedure.

Tailing cDNA seems to be less critical, although care should be taken not to let the tails become too long. The major variable in this reaction is the number of molecules of cDNA, since molecular weight estimations (e.g. from gel analysis) are only approximate.

It is advisable to monitor homopolymer addition kinetics for each new batch of enzyme using a flush-ended DNA substrate (e.g. *Alu1* cleaved pAT153) and a 3' protrusion-ended DNA substrate (e.g. *Pst1* cleaved pAT153) in pilot experiments. Taking account of the molarity of DNA 3' ends, and of deoxynucleotide precursor, the average number of bases added per end can be easily monitored by the incorporation of radioactive precursor. As far as is practicable an excess of enzyme should be used.

References

Deng, G.-R. & Wu, R. (1981) An improved procedure for utilizing terminal transferase to add homopolymers to the 3' termini of DNA. *Nucl Acid. Res.* **19**, 4173.

Jackson, D.A., Symons, R.H. & Berg, P. (1972) Biochemical method for inserting new genetic information into DNA of simian virus 40: circular SV40 DNA molecules containing lambda phage genes and the galactose operon of *E. coli*. *Proc. Natl. Acad. Sci. U.S.A.*, **69**, 2904.

Roychoudry, R., Jay, E. & Wu, R. (1976) Terminal labelling and addition of homopolymer tracts to duplex DNA fragments by terminal deoxynucleotidyl transferase. *Nucl. Acid. Res.*, **3**, 101.

Williams, J.G. (1981) The preparation and screening of a cDNA clone bank. *Genetic Engineering, Vol. 1* (ed. Williamson, R.). Academic Press, London.

Material needed

Alu1 cleaved pAT153 at 1 mg ml^{-1}

Pst1 cleaved pAT153 at 1 mg ml^{-1}

^{3}H or ^{32}P-dGTP and dCTP

Terminal deoxynucleotide transferase [~ 10 units μl^{-1} (varies from batch to batch)]

Calf thymus DNA 0·1 mg ml^{-1} in H$_2$O (carrier)

50% TCA solution

8% TCA solution

Whatman GFC filters

EtOH

0·25 M EDTA, pH 8·0

Filter washing apparatus

10 × cacodylate buffer (1.4 mK cacodylate with 0·3 M Tris base, pH to 7·6 with solid KOH)

10 mM CoCl$_2$

1 mM DTT

(a) Pilot experiment

Procedure

1 Use the analysis shown in Table A.5 to assemble reaction mixtures containing 1 pmol end of cut plasmid DNA, 2·5 µl 10× cacodylate tailing buffer, approximately 500 pmole of ^{3}H or ^{32}P dCTP or dGTP at a specific activity of 3·5 Ci mmol^{-1} and H$_2$O to 24 µl total volume.
2 Mix and place at 37°C, remove 2 µl as a zero time point.
3 Add 1 µl of enzyme (usually commercially supplied at 10 units µl^{-1}) mix thoroughly and incubate at 37°C.
4 Remove 2 µl samples at 30 s, 1 min, 2 min, 5 min, 20 min, directly into 500 µl of 50 mg ml^{-1} carrier DNA in 0·2 M sodium pyrophosphate, pH 7·0 and add 125 µl of 50% TCA. Mix and place on ice for 5–10 min.
5 Pour each sample onto a GFC filter paper in a filter tower, wash ×4 with 8% TCA and ×2 with EtOH. Dry the filters.
6 Prepare a radioactivity input control by spotting 2 µl of reaction mix directly onto a GFC filter.
7 Count all filters from steps 5 and 6 in a scintillation counter.

Table A.5 Calculation of 3′-ends available for homopolymer tailing

Part a. Assembling the reaction

pAT molecular weight is $2·5 \times 10^6$, therefore 1 pmol plasmid is 2·5 µg.
Pst1 cleaves once giving two ends, therefore 1 pmol *Pst1* cleaved 3′ ends = 1·25 µg.
Alu1 cleaves 11 times giving 22 ends, therefore 1 pmol *Alu1* cleaved 3′ ends = 0·11 µg.
For 1 pmol of 3′ ends, 25 pmol of dC should be incorporated for an average tail length of 25. Use 20-fold excess of deoxynucleotide precursor.
Therefore assemble reaction to contain 500 pmol of labelled dCTP (see Deng & Wu 1982).
For example, if ^{32}P-dCTP with a specific activity of about 1,000 Ci mmol^{-1} is used, 5 µCi will be equivalent to 5 pmol.
So add 5 µCi ^{32}P-dCTP and 1 µl of 0·5 mM dCTP to the reaction mixture. Total = 505 pmol.
If ^{3}H-dCTP is used (specific activity is usually 17 Ci mmol^{-1}) 5 µCi will be equivalent to 295 pmol. Hence add 5 µCi (remove alcohol) and 0·4 µl 0·5 mM dCTP to reaction mixture. Total = 495 pmol.

Part b. Number of bases added

Input cpm, spotted directly is say 100,000 cpm.
After 5 min 4,000 cpm are incorporated.
If all radioactivity had been incorporated 25×20 bases/end would have been added.

No. of bases/end = $25 \times 20 \times \dfrac{4,000}{100,000}$

= 20 bases/end

8 Calculate the number of bases added using the method described in Table A.5, part (b).

9 Plot a time course of the reaction in terms of the number of bases added per end. Note the time taken for addition of 30 residues.

(b) Tailing *Pst1* cleaved vector DNA (not allowed for in sample timetable, Table A.1)

5–10 µg of DNA is enough for many cloning experiments. Lightly labelled vector is stable for many months at −20°C.

1 Scale up and assemble the reaction exactly as described under pilot experiment.

2 Allow the reaction to proceed at 37°C for the desired time (based on results of pilot experiment).

Note. Kinetics of dG addition are different from those of dC addition.

3 Place reaction on ice and take a sample for counting.

4 Calculate the number of dGs added and repeat steps 2–4 if necessary.

5 When complete, stop the reaction by addition of EDTA pH 8·0 to 25 mM, phenol extract and pass through a Sephadex G50 column in TE buffer.

6 Ethanol precipitate the excluded fractions.

(c) Tailing cDNA

1 Using an estimate of 1,000 bp for the average size of cDNA (or estimate by gel analyses), calculate the number of pmol ends as in Table A.5.

2 Assemble reaction as described under pilot experiment but use enough radioactivity to achieve a specific activity of dC of 10–100× that used for cDNA synthesis (alternatively use a different isotope).

3 Proceed as for dG tailing of plasmid vector.

Experiment A.8 Annealing of oligo dC tailed double-stranded cDNA with oligo dG tailed plasmid DNA

Background

Having added homopolymeric oligo dC tails to the double-stranded cDNA preparation and oligo dG tails to *Pst*I linearized plasmid DNA, it is now possible to anneal them together to create recombinant plasmids carrying cDNA inserts (see Fig. A.6). Although such recombinant plasmids contain single-stranded regions, these are repaired *in vivo* after transformation into *Esherichia coli* to yield covalently closed plasmid species. In fact this repair is normally so accurate that the *Pst*I sites flanking the cDNA inserts are usually regenerated (Bolivar *et al.* 1977) enabling the inserts to be excised from progeny plasmids.

The annealing conditions suggested are those of Rowekamp & Firtel (1980). The annealing buffer used is 0·1 M NaCl; 10 mM Tris-HCl, pH 7.5; 0·2 mM EDTA, desirably containing approximately equimolar amounts of plasmid DNA and cDNA. Concentrations of plasmid DNA of 300–400 fmol ml^{-1} give very satisfactory results. (fmol = 10^{-15} mol).

References

Rowekamp, W. & Firtel, R.A. (1980) Isolation of developmentally regulated genes from *Dictyostelium. Dev. Biol.*, **79**, 409.

Bolivar, F., Rodriguez, R.L., Greene, P.J., Betlach, M.C., Heyneker, H.L., Boyer, H.W., Crosa, J. & Falkow, S. (1977) Construction and characterisation of new cloning vehicles. A multipurpose cloning system. (1977) *Gene*, **2**, 95

Material needed

Annealing buffer (0·1 M NaCl; 10 mM Tris-HCl, pH 7·5; 0·2 mM EDTA)

Oligo dG tailed pAT153 (> 7350 fmol ml^{-1} in annealing buffer). Approximate oligo dG tail length = 20 residues

Procedure

1 Collect the overnight ethanol precipitates of oligo dC tailed cDNA by centrifugation in the bench-top centrifuge for 15 min at 0°C.

2 Drain and dry the pellet.

3 Estimate from the amount of cDNA used in the tailing reaction what amount will now be present (in fmol) if recovery is quantitative.

4 Dissolve cDNA in a suitable volume of annealing buffer to give a concentration of 15 fmol μl^{-1}. Set up a 200 μl annealing mixture in a conical polypropylene centrifuge tube, containing approximately equimolar amounts of plasmid and cDNA (with plasmid concentration at $\simeq 350$ fmol ml^{-1}). Mix thoroughly.

5 Set up a control containing only oligo dG-tailed plasmid DNA. This can be used to assess, after transformation into *E. coli*, whether a signifiant number of plasmids with the recombinant Tc^{r}Ap^s phenotype can be obtained without cDNA being present.

6 Incubate for 5 min at 65°C, to 'unfold' single-strand tails, and then for 2 h at 44°C.

7 Switch off water bath and allow to cool slowly to room temperature overnight.

Note

1 Annealed preparations can be stored at 0°C for several weeks if necessary. The proportion of recombinants may be found to increase slightly early in storage.

2 The formation of circular 'hybrid' molecules during such annealings may be checked by electron microscopy if desired.

Section A.3

Transformation of *Escherichia coli* cells with cDNA-containing recombinant plasmid molecules and analysis of a cDNA clone-bank

Introduction

The introduction of a single plasmid molecule into a bacterial cell, by the process of transformation, is sufficient to establish a line of identical organisms, all of which contain copies of the original plasmid. Hence, the plasmid molecule can be used as a vector to achieve the clonal expansion of any DNA fragment to which it is physically linked.

Transformation of *Escherichia coli* cells with a mixture of recombinant DNA molecules consisting of vector plasmid linked to a population of DNA fragments can therefore be used to provide a bank of bacterial clones, each of which contain an amplification of a unique DNA fragment from the original population. Once established the clone bank can be screened repeatedly to identify those colonies which contain the DNA fragments of interest. Cells from these colonies can then be cultured in bulk to provide enough of the DNA fragment for further study.

The objectives of this series of experiments are:

1   Preparation of competent cells of an *E. coli* K12 strain and transformation with an 'annealed mixture' of homopolymer-tailed plasmid and cDNA molecules.

2   Identification (a) of recombinant plasmid transformant colonies by 'insertional inactivation' screening, and (b) of transformant clones carrying a specific cDNA recombinant plasmid by the Grunstein-Hogness colony hybridization procedure.

3   Rapid small scale preparation of plasmid DNA from transformant clones for physical analysis of recombinant plasmid molecules.

Experiment A.9 Transformation of competent *E. coli* cells by plasmid DNA

Background

Cells of *E. coli* K12 strains become permeable to DNA molecules when harvested during exponential growth and treated with calcium (or other divalent cation) salts at 0°C. Such cells are said to be 'competent' and are able to be transformed with circular plasmid or phage DNA molecules at a detectable frequency; 10^6–10^7 transformants μg^{-1} supercoiled plasmid DNA. (For a plasmid of molecular weight 6 Md this corresponds to one transformant cell per 10^4–10^5 plasmid molecules.) Uptake is most efficient for supercoiled circular molecules, less efficient for nicked circular molecules (10^4–10^5 transformants μg^{-1} plasmid DNA) and less efficient still for linear DNA molecules (*ca* 10^2 transformants μg^{-1} plasmid DNA) (Mandel & Higa 1970, Cohen *et al.* 1972, Cosley & Oishi 1973, Lederberg & Cohen 1974, Kushner 1978).

Transformation is achieved by mixing the plasmid DNA with the competent cells at 0°C. The mixture is briefly heated at 42°C during which time the DNA enters the cells. After uptake of DNA the cells are incubated in growth medium to allow expression of the plasmid-coded genes prior to plating on selective media. In most cases transformants containing the plasmid molecules are recovered by selecting for expression of a plasmid-coded gene conferring resistance to an antibiotic such as tetracycline. If the site at which foreign DNA is inserted into the plasmid vector molecule lies within the structural gene of a determinant conferring resistance to a second antibiotic, such as ampicillin, then recombinant plasmid transformant clones can be readily identified by screening for inactivation of that gene. Thus, *in vitro* constructed recombinant plasmids can be introduced into, detected and propagated within *E. coli* cells.

The experiment described here involves transformation of competent *E. coli* cells with an annealed mixture of complementary homopolymer-tailed cDNA molecules and DNA of the plasmid pAT153 which has been linearised at the *Pst*I site within the β-lactamase structural gene. Recovery of transformant cells is achieved by selection for expression of the intact tetracycline resistance determining gene of the vector plasmid (Fig. A.6). Recombinant plasmid transformant clones can be subsequently identified by screening for inactivation of the ampicillin-resistance determinant because of insertion of the heterologous DNA into the PstI site of the β-lactamase gene (see Experiment 10).

References

Cohen, S.N., Chang, A.C.Y. & Hsu, L. (1972) Nonchromosomal antibiotic resistance in bacteria: genetic transformation of *Escherichia coli* by R. factor DNA. *Proc Natl. Acad. Sci. U.S.A.*, **69**, 2110.

Cosloy, S.D. & Oishi, M. (1973) Genetic transformation of *Escherichia coli* K12. *Proc. Natl. Acad. Sci. U.S.A.*, **70**, 84.

Kushner, S. (1978) An improved method for transformation of *Escherichia coli* with ColEl-derived plasmids. In: *Genetic Engineering*. (Eds Boyer, H.W. & Nicosia, S.), pp. 17–23. Elsevier, Amsterdam.

Lederberg, E.M. & Cohen, S.N. (1974) Transformation of *Salmonella typhimurium* by plasmid deoxyribonucleic acid. *J. Bact.*, **119**, 1072.

Mandel, M. & Higa, A. (1970) Calcium-dependent bacteriophage DNA infection. *J. Mol. Biol.*, **53**, 159.

(a) Preparation of competent cells

Material needed

Culture of *E. coli* K12 strain suitable for use as transformation host (e.g. JA221 F$^-$, *hsd*R, *trp*EΔ, *leu*B6, *rec*A, *lac*Y)
Sterile plastic round-bottom universal bottles
Sterile 0·1 M MgCl$_2$
Sterile 0·1 M CaCl$_2$
Sterile L-broth (1% tryptone; 0·5% yeast extract, pH 7·4; 0·5% NaCl)

Procedure

1 Use a colony of the preferred *E. coli* strain, e.g. JA221, to inoculate a 5 ml L-broth culture and incubate at 37°C overnight.

2 Inoculate 1 ml of this culture to 100 ml L-broth in a 500 ml flask and incubate in a shaking waterbath until the optical density reaches $A_{650} = 0.6$.

3 Chill the culture on ice.

4 For all subsequent steps solutions and bottles that have been pre-chilled should be used.

5 Transfer the culture to four sterile round-bottom plastic universal bottles and pellet the cells at 5,000 rpm (bench top centrifuge) for 5 min at 4°C.

6 Resuspend the cell pellet in ½ vol 0·1 MgCl$_2$ (0°C) and pellet the cells at 5,000 rpm for 5 min, 4°C.

7 Resuspend the cell pellet in ½ vol 0·1 M CaCl$_2$ (0°C) and hold on ice for 20–30 min.

8 Pellet the cells at 3,000 rpm for 5 min, 4°C (Note. Good competent cell preparations often form a ring at the bottom of the tube rather than a pellet).

9 Resuspend the pellet in 1/20 vol 0·1 M $CaCl_2$ (0°C) and store on ice.

These are now competent cells.

(b) Transformation

Materials needed

Annealing-mix; this contains an annealed mixture of cDNAs with 3′ homopolymer dC extensions and the plasmid vector pAT153 (Fig. B1) which has been linearised at the *Pst*I cleavage site in the β-lactamase (Amp^R) gene and extended at the 3′ termini with homopolymer dG tails, as detailed in Section A.2, Experiment 9

Oligo dG-tailed pAT153 DNA

Supercoiled pAT153 DNA (10 μg ml^{-1})

Sterile plastic culture tubes

Sterile 1×SSC (0·15 M NaCl; 15 mM Na citrate)

Sterile L-broth

L-agar plates containing tetracycline (10 μg ml^{-1}) (L-Tet)

Procedure

Set up the following tubes:

Experimental: to each of several (e.g. 5–10) plastic tubes add 10 μl annealing mix (equivalent to 1–5 ng cDNA).

Vector control: to another tube add 10 μl dG-tailed pAT153 DNA.

Transformation control: to another tube add 0·1 μg supercoiled pAT153 DNA.

Cell control: No DNA is added to this tube.

1 To each tube add 0·1 ml 1×SSC (0°C) and 0·2 ml competent cells.

2 Mix the tube contents by gentle shaking and stand on ice for 30 min with occasional shaking.

3 Transfer the tubes to a 42°C waterbath for 2 min.

4 Return the tubes to ice for 20 min.

5 Add 2 ml L-broth to each tube and incubate at 37°C in a shaking water bath for 60–90 min.

6 For all tubes except the *Transformation control* (see below) pellet the cells at 3,000 rpm for 6 min and resuspend in 100 μl L-broth.

7 Spread the contents of each tube on one L-Tet plate.

8 *Transformation control:* spread 100 µl aliquots of the undiluted culture, ×10, ×10^2 and ×10^3 dilution on to L-Tet plates.
9 Incubate all plates overnight at 37°C.

Analysis

Experimental plates

Estimate the number of colonies on the *Experimental* plates.

Transformation frequency

Count the number of colonies on the *transformation control* plates and calculate the transformation frequency of the competent cell preparation.

Transformation frequency is expressed as transformants µg^{-1} DNA. A good transformation should give 10^6–10^7 transformants per µg supercoiled plasmid DNA.

Vector control

Check the *Vector control* plate for colonies in order to determine the background level of vector colonies in your annealing-mix transformations.

Cell control

Check the *cell control* plates for colonies in order to determine the purity of your competent cell preparation. Any colonies which arise are either non-*E. coli* contaminants or are due to contamination of the tube and/or competent cell preparation with plasmid DNA.

Experiment A.10 Analysis of recombinant plasmid transformants

Background

The *Pst*I site of the vector plasmids pAT153 and pBR322 has been commonly used for cloning cDNA preparations by the annealing of homopolymer-tailed molecules (see Section A.2). As this site lies within the plasmid β-lactamase gene, which confers ampicillin resistance to the host cell, insertion of heterologous DNA generally results in the inactivation of the gene (insertional inactivation). Transformant colonies which contain recombinant plasmids constructed in this way can be easily identified by screening for those which are sensitive to ampicillin. Note, however, that this is not found for all restriction enzyme cleavage sites within plasmid genes (e.g. Rodriguez *et al.* 1977, Villakomaroff *et al.* 1978).

Colony hybridization permits rapid screening of bacterial colonies by DNA/DNA or RNA/DNA hybridization in order to determine which contain specific DNA sequences. It is particularly useful for screening cDNA and genomic clone banks for unique DNA sequences. Colonies growing on nitrocellulose membrane filters are lysed *in situ* and the released DNA denatured and irreversibly bound to the membrane filter. The filters are then incubated with a radioactively labelled probe DNA or RNA. Hybridization occurs between the probe and the DNA of those colonies which contain homologous sequences. After washing to remove unhybridized probe, the filters are dried and exposed to X-ray photographic film. Colonies which contain DNA sequences homologous to the probe can be detected by the presence of prints on the autoradiograph.

The intensity of the autoradiograph print is a reflection of the extent of hybridization between the probe and the colony DNA. This is dependent upon several factors, including the size of the probe fragment, and the degree of homology between the probe and the hybridizing DNA and the relative abundance of the hybridizing molecules in the labelled probe population. Similarly, the conditions under which the hybridization and subsequent washes are carried out can affect the extent of hybridization. In general the stringency of the hybridization conditions can be increased by either decreasing the salt concentration or increasing the hybridization temperature. The reader should consult the excellent reviews by

Grunstein and Wallis (1979) and Williams (1981) for a detailed discussion on these aspects of colony hybridization, and refer also to Experiment 2(f) and (g), Section B of this manual.

The method detailed here is a modification by Young and Hogness (1977) of the original procedure devised by Grunstein and Hogness (1975). This procedure is suitable for use with labelled DNA probes made by either cDNA synthesis (Experiment 4) or nick translation (Experiment 3, Section B) and employs low stringency hybridization conditions which allow the detection of both weak and strong hybridization responses.

The original method has been modified so that it is now possible to screen phage plaques by a similar method (Benton & Davis 1977). Other adaptations have allowed the screening of yeast transformants (Hinnen et al. 1978) and animal cell SV40-induced plaques (Villarreal & Berg 1977) for unique DNA sequences and both bacterial colonies (Villakomaroff et al. 1978) and phage plaques (Sanzey et al. 1976) for the expression of functional gene products by use of specific antibodies.

References

Benton, W.D. & Davies, R.W. (1977) Screening λgt recombinant clones by hybridization to single plaques *in situ*. *Science*, **196**, 180.

Grunstein, M. & Hogness, D.S. (1975) Colony Hybridization: A method for the identification of cloned DNAs that contain a specific gene. *Proc. Natl. Acad. Sci. U.S.A.*, **72**, 3961.

Grunstein, M. & Wallis, J. (1979) Colony hybridization. *Meth. Enzym.*, **68**, 379.

Hinnen, A., Hicks, J.B. & Fink, G.R. (1978) Transformation of yeast. *Proc. Natl. Acad. Sci. U.S.A.*, **75**, 1929.

Rodriguez, R.L., Tait, R., Shine, J., Bolivar, H., Heyneker, H., Betlach, M. & Boyer, H.W. (1977) Characterization of tetracycline and ampicillin resistance plasmid cloning vehicles. In: *Molecular Cloning of Recombinant DNA*. Vol. 13. Miami Winter Symposium (Eds. Scott, W.A. & Werner, R.), pp. 73–84. Academic Press, New York and London.

Sanzey, B., Mercereau, O., Ternynck, T. & Kourilsky, P. (1978) Methods of identification of recombinants of phage λ. *Proc. Natl. Acad. Sci. U.S.A.*, **73**, 3394.

Villakomaroff, L., Efstratiadis, A., Broome, S., Lomedico, P., Tizard, R., Naber, S., Chick, W.L. & Gilbert, W. (1978) A bacterial clone synthesizing proinsulin. *Proc. Natl. Acad. Sci. U.S.A.*, **75**, 3727.

Villarreal, L.P. & Berg, P. (1977) Hybridization *in situ* of SV40 plaques: detection of recombinant SV40 virus carrying specific sequences of non viral DNA. *Science*, **196**, 183.

Williams, J.G. (1981) The preparation and screening of a cDNA clone bank. In: *Genetic Engineering*. Vol. 1. (Ed. Williamson, R.), pp. 1–59. Academic Press, London.

Young, M.W. & Hogness, D.S. (1977) A new approach for identifying and mapping structural genes in *Drosophila melanogaster*. In: *Molecular Approaches to Eukaryotic Genetic Systems*. ICN-UCLA Symposia on Molecular and Cellular Biology VIII. (Eds. Wilcox, G., Maclean, J. & Fox, C.F., pp. 315–351. Academic Press, New York.

(a) Screening transformant colonies for recombinant plasmid clones

Materials needed

Nitrocellulose membranes (85 mm diameter)
L-Tet plates (10 μg ml^{-1} tetracycline)
L-Ap plates (25 μg ml^{-1} ampicillin)
Sterile toothpicks
'Marker' grids (Fig. A.7)

Procedure

1 Using sterilized forceps transfer a nitrocellulose membrane filter onto an L-Tet plate.
2 Mark the filter, a second L-Tet plate and an L-Ap plate with an orientation mark and an identification code.
3 Using sterile toothpicks (one per colony) transfer tetracycline resistant transformant colonies from the *Experimental* plates (see Experiment A.9) firstly to the L-Ap plate, next to the filter on the L-Tet plate and finally to the extra L-Tet plate. Arrange the colonies in an orderly array using the marker grid as a guide and aim to get about 100 colonies on a plate. Pick *at least* 300 colonies; the more the better.
4 As controls, pick some (20–30) colonies from the *Vector control plate* and a few colonies of the pAT153 transformants (*Transformation control*).
5 Incubate the plates at 37°C overnight.

Analysis of data

Compare the number of colonies on the L-Ap and L-Tet plates and calculate the percentage of recombinant plasmid transformants. For a carefully prepared cDNA/Plasmid 'annealing mixture', these should exceed 80% of the total transformants. The colonies growing on the nitrocellulose membrane filter on the L-Tet plate, can be used for a colony hybridization screening experiment to determine which contain specific cDNA recombinant plasmids. The colonies growing on the extra L-Tet plate can be kept as a stock and used to inoculate small liquid cultures for rapid microscale analysis of the recombinant plasmid molecules (Experiment A.11).

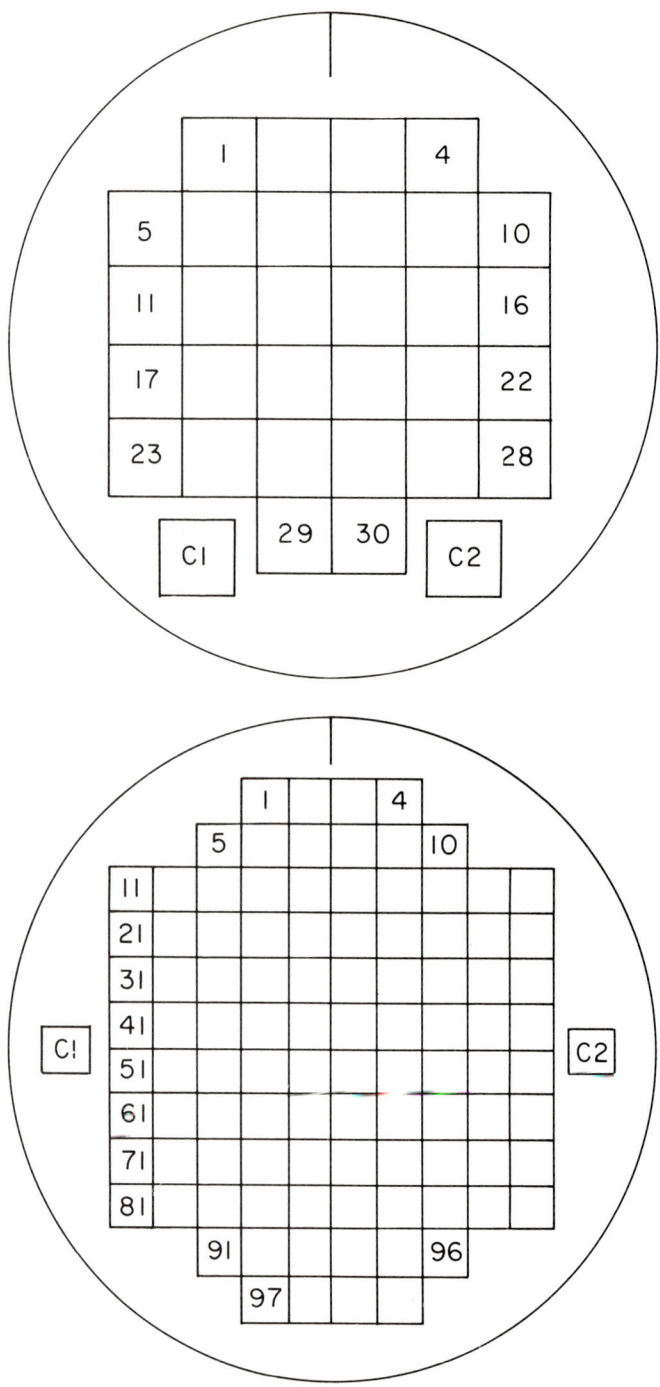

Fig. A.7 *Marker grids.* Photocopies of these grids can be stuck into petri dish lids for use as a location guide underneath agar plates when tooth-picking bacterial or phage clones.

(b) Colony hybridization to identify clones carrying specific cDNA recombinant plasmids

Material needed

DNA binding

Filter/blotting paper
Dishes
0·5 M NaOH
1·0 M Tris-HCl, pH 7·4
1·5 M NaCl, 0·5 M Tris-HCl, pH 7·4
Vacuum desiccator

Hybridization with probe DNA

^{32}P-radioactively labelled DNA probe. This can be prepared by nick translation (Experiment B.2) or by cDNA synthesis (Experiment A.4)
Plastic petri dishes
PVC tape
5× Denhardt solution (0·1% bovine serum albumin, 0·1% Ficoll, mol. wt. 400,000, 0·1% Polyvinyl pyrrolidone, mol. wt. 40,000) in 6× SSC—prewarmed to 65°C and degassed
Plastic 'lunch-boxes'
0·1% SDS in 3× SSC prewarmed to 65°C and degassed
Whatman 3 MM filter paper
Clingfilm
X-ray film
X-ray cassette

(i) Cell lysis, DNA-denaturation and DNA-binding

Procedure

1 Prepare four dishes (a–d) by putting in them 6–10 layers of filter/blotting paper soaked with appropriate solutions (a) 0·5 M NaOH, (b) 1·0 M Tris-HCl, pH 7·4, (c) 1·0 M Tris-HCl, pH 7·4, (d) 1·5 M NaCl, 0·5 M Tris-HCl, pH 7·4.
2 Remove the nitrocellulose filters with the transformant colonies from the L-Tet plate and place, *colony side up,* on the 0·5 M NaOH saturated filter paper for 7–8 min. Do not allow any liquid onto the upper surface or the colonies may merge.
3 Transfer the filters to the 1·0 M Tris-HCl, pH 7·4 saturated filter paper for 2 min.
4 Transfer the filters to the second dish of 1·0 M Tris-HCl, pH 7·4, for 2 min.

5 Transfer the filters to the 1·5 M NaCl, 0·5 M Tris-HCl, pH 7·4, saturated filter paper for 4 min.
6 Dry the filters by placing them on dry filter paper in a 60°C oven for ~ 15 min.
7 Bake the filters in an evacuated vacuum desiccator at 80°C for 2 h.

Once prepared, the filters can be stored individually in filter paper envelopes in a dry place until required for hybridization.

(ii) Hybridization with a radioactively labelled probe DNA

Procedure

1 Immerse the filters with the immobilized DNA in pre-warmed de-gassed 5× Denhardt solution in a petri dish (three filters per dish) and place at 65°C for 3–4 h.
2 Remove the filters and place three or four to a dish into plastic petri dishes containing 15–20 ml 5× Denhardt solution (prewarmed and de-gassed).
3 *Boil* the radioactively labelled hybridization probe DNA for 3–5 min at 100°C in a waterbath.
4 Add probe to the dishes containing the filters. (Try to add approximately $1-3 \times 10^6$ cpm per dish, ideally 10^6 cpm filter^{-1}; if necessary use all of the probe).
5 Seal the lids of the petri dishes with PVC tape and shake gently to mix the probe into the solution, and to ensure that all of the filters are in contact with the radioactive solution.
6 Place the petri dishes in a plastic lunch-box containing two or three layers of filter paper soaked in water and seal the lid with PVC tape.
7 Place the box of dishes in the 65°C incubator and allow the hybridization to proceed overnight.
Next day proceed to the washing of filters.
8 Transfer the filters from the hybridization mixes to fresh dishes containing prewarmed, de-gassed 0·1% SDS in 3× SSC.
9 Place the dishes in the plastic lunch-box at 65°C; leave for 45–60 min with occasional shaking.
10 During the next 4 h repeat this washing procedure four more times.
11 Dry the filters on a sheet of filter paper in a 37°C incubator.
12 Mount the filters onto a sheet of Whatman 3 MM paper and scan them with a hand monitor to see if there are any localized 'hot spots' of activity.

13 Cover the filters with Clingfilm and expose them to an X-ray film, with an intensifying screen at −70°C.

14 Allow to expose as long as possible (overnight) and develop the autoradiograph (Appendix C.III).

15 Identify those colonies which contained cDNA homologous to the specific probe.

Experiment A.11 Analysis of the plasmids present in recombinant clones

Background

Several methods have been developed in recent years for rapid isolation of bacterial plasmid DNA. Some can be performed on a microscale from small (0·5–5·0 ml) liquid cultures or single colonies, to yield enough plasmid DNA for a few gel electrophoresis analyses. (Barnes 1977, Meagher *et al.* 1977, Telford *et al.* 1977, Eckhardt 1978, Birnboim & Doly 1979, Klein *et al.* 1980, Holmes & Quigley 1981, Kado & Liu 1981). However, not every method is suitable for all *E. coli* strains, and not all give DNA preparations suitable for digestion with restriction endoncleases. Both the 'alkaline extraction' method of Birnboim and Doly (1979) and the 'acid-phenol' extraction of Klein *et al.* (1980) give good quality preparations but involve several manipulations. The method described here is based upon that developed by Holmes and Quigley (1981) and allows rapid isolation of good quality plasmid DNA on a microscale.

Whilst most procedures are successful with small plasmids it should be noted that very few methods work well with large plasmids. Screening for large plasmids is still a problem therefore and usually it is necessary to resort to a full scale CsCl-ethidium bromide preparation to identify the presence of such plasmids. One recent report suggests that an alternative screening method (Kado & Liu 1981) can cope with these plasmids and preliminary results with the ColV plasmid in Leicester are consistent with this.

In practice, particularly when screening cDNA recombinant clones produced by the methods described in Experiments A.4–A.10, it is often useful to run two of these rapid screening procedures. Firstly single colonies lysed with SDS (e.g. Barnes 1977, Appendix A.VI) can be used to rapidly and simultaneously 'size-screen' many colonies in order to determine which contain the largest recombinant plasmids. Once identified these colonies can be used for a second rapid lysis procedure, e.g. Holmes and Quigley (1981) or Birnboim and Doly (1980), to give a small quantity of plasmid DNA suitable for restriction analysis in order to determine which clones contain the particular plasmids of interest.

References

Barnes, W.M. (1977) Plasmid detection and sizing in single colony lysates. *Science*, **195**, 393.

Birnboim, H.C. & Doly, J. (1979) A rapid alkaline extraction procedure for screening recombinant plasmid DNA. *Nucl. Acid. Res.*, **7**, 1513.

Eckhardt, T. (1978) A rapid method for the identification of plasmid deoxyribonucleic acid in bacteria. *Plasmid*, **1**, 584.

Holmes, D.S. & Quigley, M. (1981) A rapid boiling method for the preparation of bacterial plasmids. *Analyt. Biochem.*, **114**, 193.

Kado, C.I. & Liu, S.T. (1981) Rapid procedure for detection and isolation of large and small plasmids. *J. Bacteriol.*, **145**, 1365.

Klein, R.D., Selsing, E. & Wells, R.D. (1980) A rapid microscale technique for isolation of recombinant plasmid DNA suitable for restriction enzyme analysis. *Plasmid*, **3**, 88.

Meagher, R.B., Tait, R.C., Betlach, M. & Boyer, H.W. (1977) Protein expression in *E. coli* minicells by recombinant plasmids. *Cell*, **10**, 521.

Telford, J., Boseley, P., Schaffner, W. & Birnsteil, M. (1977) Novel screening procedure for recombinant plasmids. *Science*, **195**, 391.

(a) Rapid microscale plasmid preparation

Materials needed

Sterile culture tubes
Sterile L-broth
STET buffer (8% sucrose; 5% Triton ×100; 50 mM EDTA; 50 mM Tris-HCl, pH 8·0)
Lysozyme 10 mg ml^{-1}, freshly prepared
Isopropanol
0·3 M Na acetate, pH 6·5
Ethanol (absolute) $-20°C$

Procedure

1 Select 10 colonies which contain recombinant plasmids (e.g. ampicillin-sensitive transformant colonies of Experiment A.9/A.10) and use them to inoculate 1 ml L-broth cultures. A selection for the presence of the plasmid (e.g. an antibiotic) can be included if desired. As controls, inoculate another tube with a colony containing the vector plasmid (e.g. pAT153 from the *Transformation Control* plate of Experiment A.10) and a final tube with a colony of the plasmid-free host strain (e.g. JA221 from Experiment A.9).

2 Incubate the cultures at 37°C overnight.

Next day

All manipulations of the lysis procedure should be performed at room temperature unless otherwise stated.

1 Transfer 0·5 ml of each 1 ml overnight culture to a small Eppendorf tube (0·5 ml).

2 Pellet the cells for 20 s at 12,000 g (Eppendorf microfuge) and remove the supernatant.
3 Resuspend the cell pellet in 25 µl STET buffer by vortex mixing.
4 Add 2 µl lysozyme solution and vortex mix.
5 Place the tube in a boiling water bath for 40 s.
6 Remove the tubes and immediately centrifuge in the Eppendorf microfuge for 10 min (room temperature).
7 Carefully remove the supernatant fluid (20 µl) overlying the slightly gelatinous pellet to a fresh tube. (A 25 µl or 50 µl capillary-pipette with a screw micropipettor may be useful.) Discard the pellet.
8 Add an equal volume of isopropanol (20 µl) and precipitate the nucleic acids at −70°C (ethanol/dry-ice bath) for 10 min.
9 Pellet the nucleic acids by centrifugation at 12,000 g at 4°C for 10 min (Eppendorf microfuge). Carefully decant the supernatant.
10 Resuspend the pellet in 25 µl 0·3 M Na acetate, add 75 µl absolute ethanol and re-precipitate the nucleic acids at −70°C for 10 min.
11 Pellet the nucleic acids by centrifugation at 12,000 g at 4°C for 10 min (Eppendorf microfuge). Very carefully, without disturbing the pellet, rinse the tube with 100 µl 70% ethanol (−20°C) and spin at 12,000 g for a few minutes in order to ensure that the pellet is firm.

Carefully decant the supernatant and dry the pellet under vacuum for 5–10 min.
Resuspend the pellet in 10 µl distilled water.

The plasmid DNA can now be analysed by gel electrophoresis, either undigested or after digestion with a restriction endonuclease. This procedure generally gives 0·1–0·5 µg plasmid DNA which is sufficient for one or two enzyme digests. Note that the preparation contains large amounts of RNA. If small DNA fragments are to be visualized it is necessary to treat the digested DNA with RNAase. A rapid microscale procedure for analysing plasmid DNAs prepared in this way is outlined below.

(b) Rapid analysis of microscale plasmid DNA preparations

Materials needed

10× restriction endonuclease buffer (use suppliers' suggested conditions for the appropriate enzyme)
Restriction endonuclease (1–2 units µl^{-1})

Pancreatic RNase (1 mg ml^{-1} in 5 mM Tris-HCl, pH 7.5; previously heated to 100°C for 10 min and slow cooled)
10× loading dye (50% glycerol; 0.2 M EDTA; 50 μg ml^{-1} bromophenol blue)

Procedure

1 To clean sterile small microfuge tubes add 5–10 μl plasmid DNA preparation, 0.5–1 μl 10× restriction endonuclease buffer and 1–2 units restriction endonuclease.
2 Incubate at 37°C for 30–60 min.
3 Add 1 μl pancreatic RNAase to each tube and continue the incubation at 37°C for 20–30 min.
4 Add 1 μl loading dye, vortex mix and spin the tubes in a microfuge for 10 s.
5 Load the samples to 1% agarose gel (made on a microscope slide; see below) and electrophorese until the marker dye has run three-quarters of the gel length.
6 Soak the gel in ethidium bromide (0.5 μg ml^{-1}) for 10 min to stain the DNA bands and visualize under ultraviolet light.

(c) Microscope slide agarose gel

Materials needed

10 ml 1% agarose in electrophoresis buffer (40 mM Tris; 20 mM Na acetate; 2 mM EDTA, pH 7.8)
3″×2″ microscope slide
Ethidium bromide solution (0.5 g/ml^{-1})
Care! Ethidium bromide is a mutagen.

Procedure

1 Using 'fold-back' clips or plasticine, position a comb with small teeth (4×1 mm) so that it is about one-quarter distance along and 1 mm above a microscope slide.
2 Dissolve the agarose in electrophoresis buffer by boiling and cool until 'hand-hot', i.e. about to gel.
3 With a pipette carefully, but quickly, transfer the agarose onto the microscope slide (10 ml for a slide) so that it is retained by surface tension.
4 Allow the agarose to gel (10–20 min).
5 With a Pasteur pipette place a small amount of electrophoresis buffer onto the gel around the comb and very carefully remove the comb. The buffer should lubricate the teeth but take care not to crack either the gel itself or the agarose film at the base of the wells under the comb.

6 Submerge the gel on the slide in electrophoresis buffer in a horizontal gel apparatus.

7 Carefully load the samples into the gel wells (5–6 µl per well). A 5 or 10 µl capillary pipette in a micropipetter is useful for this.

8 Electrophorese the samples until the marker dye has run three-quarters the length of the gel. (If a voltage of 100 V is applied, the gel will generally have run sufficiently far in 30–45 min.)

9 Remove the gel and immerse in ethidium bromide solution for 10 min to stain the DNA bands and then visualize under ultraviolet light.

Appendix A.I Preparation and use of micrococcal nuclease-treated reticulocyte lysates

Induction of anaemia and preparation of reticulocyte lysates from rabbits

Rabbits

New Zealand White of weight 2–2·5 kg. Use three or four in order to obtain at least one active lysate.

Anaemia

Six daily injections (subcutaneous, beneath skin behind back of head) of 0·25 ml of 2·5% acetylphenylhydrazine (Sigma, prepared in sterile distilled H_2O) per kg body weight, i.e. 0·75 ml for 3 kg rabbit. This concentration of acetylphenylhydrazine is at the limit of solubility, so warm the solution carefully before injection. Store the solution at $-20°C$.

Reticulocytosis can be checked by methylene blue smear (although this is not routinely done).

Lysate preparation

Blood is removed by cardiac puncture after heavy ether anaesthesia, withdrawing into plastic 50 ml syringes previously coated with 5 mg/ml heparin, and expelled into Ehrlenmeyer flasks on ice containing more heparin. Cells are washed thoroughly (at least $4\times$) in 140 mM NaCl, 1·5 mM Mg acetate, 5 mM KCl in order to remove heparin (a potent inhibitor of protein synthesis). Remove the 'buffy coat' of white cells by careful aspiration. Estimate the packed cell volume. Add equal volume cold sterile distilled water and disperse the cells. Centrifuge immediately at 15 000 g in sterile Corex glass tubes for 20 min.

Remove supernatant carefully and quick freeze immediately in liquid N_2 in 1·0 ml aliquots. The lysate can be stored in the long term (1 year) in liquid N_2 or for shorter periods (months) at $-70°C$.

Micrococcal nuclease treatment to reduce background globin synthesis

1   Rapidly thaw out a 1·0 ml aliquot of lysate in the presence of 40 μl 1 mM stock haemin solution to prevent induction of the haemin-controlled repressor of protein synthesis. See below for details of haemin stock solution.
2   Add 10μl 0·1 M $CaCl_2$ (1 mM final). 10μl micrococcal nuclease solution (10^4 units/ml H_2O, stored $-20°C$).
3   Incubate in 100 μl aliquots at 20°C for 10–15 min. (Check the time course of the reaction in a 'pilot' experiment, assaying background and mRNA-directed protein synthesis with and without a standard mRNA such as TMV RNA.)
4   Inhibit the nuclease by addition of EGTA to 4 mM final concentration from a 0·75 M sterile stock solution.
5   Freeze the aliquots and store at $-70°C$.

Preparation of haemin stock solution

Dissolve 6·5 mg of haemin in 0·25 ml 1 M KOH.
Add following sequentially:
   0·5 ml 0·2 M Tris-HCl (pH 7·8),
   8·9 ml ethylene glycol,
   0·19 ml 1 M HCl,
   0·04 ml H_2O.

This gives a 1 mM solution in 90% ethylene glycol/10 mM Tris. It can be stored indefinitely at $-20°C$.

It is useful to prepare a blank solvent preparation without haemin simultaneously.

Additional notes for use of self-prepared lysates

Commercially-supplied lysates usually contain added 'cold' amino acids and an energy-generating system. For self-prepared lysates it is necessary to add, in a 25 μl reaction mixture:

   1·0 μl 19 amino acids (minus methionine); 1 mM each in H_2O (stored $-20°C$)
   1·0 μl energy-generating system (25 mM ATP, 25 mM GTP, 0·2 M creatine phosphate and 5 mg ml^{-1} creatine phosphokinase. All in H_2O and stored at $-70°C$
   1·25 μl 2 M KCl
   1·0 μl 25 mM Hepes-KOH, pH 7·2, 25 mM Mg acetate
   1·0 μl calf-liver tRNA (1 mg ml^{-1})

Amino acid incorporation into protein may be conveniently assayed by the method described by Pelham and Jackson (1976).

Reference

Pelham, H.R.B. & Jackson, R.J. (1976) An efficient mRNA-dependent translation system from reticulocyte lysates. *Eur. J. Biochem.*, **67**, 247.

Appendix A.II Alternative method for second-strand cDNA synthesis

Procedure

1 Purify first-strand cDNA/RNA hybrid by Sephadex G100 gel filtration and precipitate with ethanol.

2 Redissolve pellet in 0·5 mM EDTA at a concentration of 80 μg cDNA ml^{-1}, heat to 100°C for 2 min, chill rapidly. Assemble 200 μl reaction mixture as follows:

250 mM Hepes buffer, pH 6·9	80 μl
cDNA (as above)	25 μl
1 M MgCl$_2$	1 μl
1 M KCl	14 μl
5 mM (each) dNTPs	20 μl
1 M dithiothreitol	4 μl
^{3}H or ^{32}P-dNTP	5–10 μCi
DNA polymerase (4 units ml^{-1}) Klenow fragment	10 μl
H$_2$O	200 μl

3 Incubate at 13°C for 7 h.

4 Monitor second-strand synthesis by TCA-precipitation of incorporated isotope.

Appendix A.III A procedure for characterizing preparations of S1 nuclease

Background

S1 nuclease preparations should be monitored to determine their capacity to degrade single-stranded cDNA in the presence of double-stranded DNA restriction endonuclease fragments of similar size to cDNA preparations (e.g. a range of 200–1,500 base pairs). Single-stranded cDNA degradation can be followed by TCA precipitation over the time course of the reaction. Nucleolytic activity can be detected by polyacrylamide gel electrophoresis of the DNA. Endonucleolytic cleavage will be most apparent as a reduction in the molecular weight of large restriction fragments; exonucleolytic attack will be more easily detected as a reduction in size of the smaller restriction fragments.

Materials needed

0·5 µg ^{32}P-labelled single-stranded cDNA (e.g. from globin mRNA)
2·5 µg *Taq*I digested pBR322 DNA, phenol extracted, ether extracted, ethanol precipitated and dissolved at 1 µg 15 µl^{-1} in TE buffer.
5×S1 buffer (1·4 M NaCl; 22·5 mM $ZnCl_2$; 150 mM sodium acetate, pH 4·5)
S1 nuclease
10% SDS
0·4 M EDTA, pH 8·0
Tris-saturated redistilled phenol

Procedure

1 Mix the cDNA and plasmid DNA preparations so that about 80% of the DNA is double-stranded.
2 Add 240 µl of 5×S1 buffer and water to 1·2 ml.
3 Remove zero time sample of 25 µl for TCA precipitation: mix with 500 µl 50 µg ml^{-1} salmon sperm DNA in 0·2 M sodium pyrophosphate (pH 7·0), add 150 µl 50% ice cold TCA, stand on ice 15 min, filter through a Whatman GF/C filter under vacuum, wash filter with 5×10 ml 5% ice cold TCA and 2×5 ml ethanol, dry filter, add filter to 5–10 ml scintillation fluid and count.

4 Remove zero time sample of 175 μl for electrophoretic analysis: mix with 300 μl 0·1 M Tris (pH 7·5) and 500 μl phenol and centrifuge. Remove aqueous phase. Ether extract. Ethanol precipitate. Dissolve DNA and analyse on 5% polyacrylamide gel.
5 Add 375 units S1 nuclease to remaining cDNA/plasmid DNA mixture and incubate at 37°C.
6 Withdraw 25 μl and 175 μl aliquots to monitor TCA precipitable single-stranded cDNA and restriction endonuclease fragment character at e.g. 5-min, 15-min, 30-min and 60-min intervals.

Interpretation

Analysis of results allows:
1 Determination of the time course of single-stranded cDNA degradation.
2 Assessment of any endonucleolytic digestion of higher molecular weight restriction fragments.
3 Assessment of any exonucleolytic digestion of lower molecular weight restriction fragments.
4 Determination of the time course of nucleolytic degradation of double stranded DNA.

Appendix A.IV Analysis of products of cDNA synthesis by denaturing polyacrylamide gel electrophoresis

This procedure describes a rapid system for analysis of the products arising from first-strand cDNA synthesis (Experiment 4), second-strand cDNA synthesis (Experiment 5) and treatment of double-stranded cDNA with S1 nuclease (Experiment 6). The combination of urea and heat in the gel generated by the relatively high current ensures that the samples are completely denatured and migrate through the gel according to their chain length (i.e. molecular weight).

Material needed

Vertical slab gel electrophoresis apparatus suitable for running a 20 cm gel, 4 cm wide, 1·5 mm thick, Scotch tape.
Acrylamide ⎫ electrophoresis grade
Bis-acrylamide ⎭
Urea (A.R.)
10×TBE buffer (0·9 M Tris base; 0·9 M boric acid, pH 8·3; 25 mM EDTA)
Whatman No. 1 filter paper
Ammonium persulphate 10% solution
TEMED
Formamide-dye mix (90% Formamide A.R., 25 mM EDTA, 0·5% bromophenol blue, 0·5% xylene cyanol FF. (Note. The Formamide should be deionized by stirring with Amberlite MB-1 resin before use)
^{32}P-labelled marker DNA fragments (e.g. *Hae*III digested ΦX174 DNA)
Clingfilm

Procedure

Preparation of the gel

1 Thoroughly clean the glass plates and assemble the apparatus, using 1·5 mm gel spacers (lightly greased with vaseline) and seal the edges with electrical tape.
2 To clean beaker add: 2·5 g acrylamide, 0·083 g bis-acrylamide, 21 g urea, 2·5 ml 10×TBE buffer and distilled water to approximately 40 ml.

3 Dissolve the solids, heating in a 37°C waterbath if necessary.
4 Make the volume up to 50 ml with distilled water and filter the solution through Whatman No. 1 paper.
5 De-gas the solution.
6 Add 100 µl 10% ammonium persulphate.
7 Add 50–100 µl TEMED and swirl to mix.
8 Immediately pour the gel and insert the comb.
9 Leave the gel to polymerize for approximately 30 min.
10 Carefully remove the comb and immediately rinse out the sample wells to avoid secondary polymerization and slot distortion.
11 Remove the tape and spacer from the bottom of the plates and mount the gel sandwich onto the apparatus.
12 Fill the reservoir with $0.5 \times$ TBE buffer.

Appendix A.V Additional notes and trouble shooting for Experiment A.9

Transformation

Although described for use with an annealed mixture of plasmid and cDNA molecules, the transformation procedure described is equally suitable for use with ligated DNA mixtures and supercoiled plasmid DNA preparations. The higher transformation efficiency of these types of DNA preparations requires plating a more dilute aliquot of the transformation mixture in order to obtain single cell transformant clones.

When transforming annealed mixtures of homopolymer-tailed plasmid vector and cDNA fragments it is important that a control transformation be performed with the tailed plasmid DNA alone. This allows estimation of the number of uncut circular molecules present in the plasmid population. It is these molecules which will primarily constitute the background of non-recombinant plasmid transformants recovered from the experiment, because of their higher transformation efficiency.

Several factors influence the choice of *E. coli* K12 strain to be used as a transformation host. As DNA to be cloned is often of non-*E. coli* origin it is desirable to use a strain which is deficient in the natural system by which *E. coli* cells recognise and degrade foreign DNA; i.e. a restrictionless (*hsd*R or *hsd*S) mutant. Host-mediated restriction of foreign DNA can reduce the overall transformation frequency by several orders of magnitude. Similarly, in order to preclude structural rearrangement of the recombinant molecules which might occur during clonal amplification of DNA fragments it is prudent to use a host deficient in normal homologous recombination functions (Rec$^-$).

The presence of even very low concentrations of detergents markedly affects the efficiency of uptake of DNA by competent cells. If glass tubes and bottles are used for the preparation and transformation of competent cells it is particularly important that they are extensively rinsed with distilled water after washing. Treatment of glass tubes and bottles with a siliconising solution (e.g. Sigmacote, Sigma) prevents possible irreversible binding of the transforming DNA to the vessel walls. The use of sterile, disposable plastic ware circumvents both of these possible problems. Note, too, that the choice of media used for growth of the culture can affect the competence of the cells. Experience at Leicester has shown that good competent cell preparations can be made from L-broth cultures but not from Oxoid Nutrient-Broth cultures.

Use of disabled hosts

Because of the conjectured hazards in the propagation of recombinant DNA molecules in bacterial cells, several laboratories have constructed strains which have dramatically reduced viability in the natural environment. These 'disabled' strains such as X1776 (Curtiss *et al.* 1977) and MRC8 (GMAG 1980) therefore provide a form of biological containment because they contain mutations rendering them unable to survive outside the artificial environment of a laboratory flask. Unfortunately these strains are often found to be so enfeebled that they are particularly difficult to work with. In this context it should be mentioned that introduction of the *recA* mutation which eliminates host recombination functions also affords a certain level of biological containment. This mutation also renders the strain deficient in post-replication DNA repair functions, thereby causing it to be sensitive to UV light (sunlight) and to have a reduced viability.

Within the Leicester University laboratories the preferred strains for use as transformation hosts are JA221 (Clark & Carbon 1978) and HB101 (Boyer & Roulland-Dussoix 1969). Both strains are restriction and recombination deficient, have a high viability and give good transformation frequencies $\geq 10^6$ transformants μg^{-1} supercoiled plasmid DNA. When used in conjunction with a non-mobilizable plasmid vector such as pAT153 (Twigg & Sherratt 1980) they provide a high level of biological containment.

Storage of clones

Once recombinant clones have been produced it is advisable to transfer them to a long-term storage system. Individual recombinant clones can be stored as frozen glycerol cultures at $-20°C$ or $-70°C$ (Miller 1972, Williams & Lloyd 1979). For a bank of several thousand clones this becomes a time consuming and tedious exercise. An alternative procedure has been suggested by Hanahan and Meselson (1980) in which recombinant clones are grown as microcolonies on a nitrocellulose membrane filter which is then stored, with the colonies remaining *in situ*, frozen in glycerol. In this way several thousand clones can be accommodated on a few filters in a minimum of freezer space.

References

Boyer, H.W. & Roulland-Dussoix, D. (1969) A complementation analysis of the restriction and modification of DNA in *Escherichia coli*. *J. Mol. Biol.*, **41**, 459.

Clarke, L. & Carbon, J. (1978) Functional expression of cloned yeast DNA in *Escherichia coli*: specific complementation of arginino succinate lyase (*arg* H) mutations. *J. Mol. Biol.*, **120**, 517.

Curtiss, R., Inouye, M., Pereira, P., Hsu, C.J., Alexander, L. & Rock, L. (1977) Construction and use of safer bacterial host strains for recombinant DNA research. In: *Molecular Cloning of Recombinant DNA*. Miami Winter Symp. Vol. 13. (Eds. Scott, W.A. & Werner, R.), pp. 99–111. Academic Press, New York and London.

G.M.A.G. (1980) Host/Vector Systems. Genetic Manipulation Advisory Group, Note 9, Supplement 1, January (1980) Medical Research Council, 20 Park Crescent, London W1N 4AL.

Hanahan, D. & Meselson, M. (1980) Plasmid screening at high colony density. *Gene*, **10**, 63.

Miller, J.H. (1972) *Experiments in Molecular Genetics*. Cold Spring Harbor Laboratory, Cold Spring Harbor, N.Y., U.S.A.

Twigg, A. & Sherratt, D. (1980) Transcomplementable copy number mutants of plasmid Col E1. *Nature*, **283**, 216.

Williams, J.G. & Lloyd, M.M. (1979) Changes in the abundance of polyadenylated RNA during slime mould development measured using cloned molecular hybridization probes. *J. Mol. Biol.*, **129**, 19.

Appendix A.VI Single colony SDS lysate

1 Resuspend a single colony in three or four drops TE buffer in a microfuge tube.
2 Add two drops 10% SDS solution.
3 Hold at room temperature or warm at 37°C for a few minutes until cell lysis occurs.
4 Clear the lysate by spinning at 12,000 g for 12 min at room temperature (Eppendorf microfuge).
5 Remove the supernatant and mix with loading dye.
6 Load the sample to an agarose gel for electrophoresis of the plasmid DNAs (Fig. A.8).

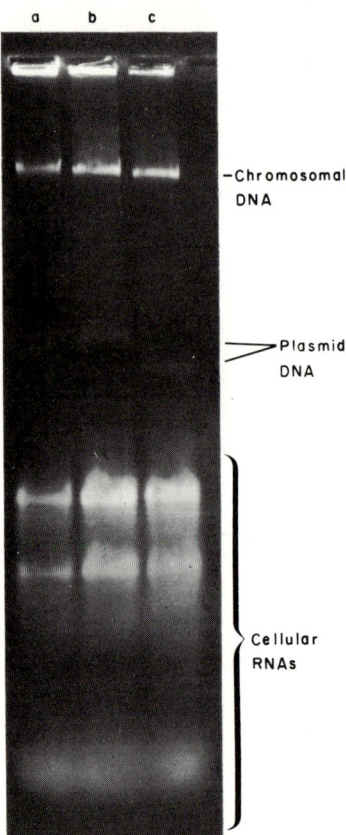

Fig. A.8 An agarose gel of plasmid DNAs prepared as above from three *E. coli* strains. (a) *E. coli* JA221 containing no plasmid; (b) *E. coli* JA 221 containing the vector plasmid pAT153; (c) *E. coli* JA221 containing a cDNA-pAT153 recombinant plasmid.

Appendix A.VII Preparation of plasmid DNA

Several methods exist for the preparation of plasmid DNA. Most vary only in detail and the three given here have been used successfully in Leicester. Advantage can be taken of the fact that in the case of many small plasmids, copy number can be amplified several hundred-fold by growth of the host in the presence of an antibiotic which inhibits protein synthesis. Under these conditions replication of the chromosome is blocked but plasmid replication continues. This occurs in the case of ColEI and its derivatives (e.g. pBR322) and some other high copy number plasmids, but not generally in the case of low copy number plasmids. Some methods appear better suited than others for preparing large plasmids but it is frequently a matter of trial and error, or even using alternative host strains, in order to ensure good recoveries.

Method 1 (Triton lysis)

1 Grow plasmid-containing strain with vigorous aeration in L-broth (250 ml) containing appropriate antibiotic selection to $A_{650} = 0.8$. If the plasmid can be amplified add chloramphenicol to a final concentration of 100–200 μg ml^{-1}. Continue the incubation overnight.

2 Next day collect the bacteria by centrifugation for 10 min at 6000 rpm at 4°C and resuspend the pellet in 3 ml ice-cold 25% sucrose in 50 mM Tris-HCl, pH 8.0.

3 Place the cell suspension on ice and add 0.5 ml of 10 mg/ml lysozyme in 50 mM Tris-HCl, pH 8.0. Mix and incubate on ice for 15 min.

4 Add 1.0 ml 0.25 M EDTA, pH 8.0, mix and continue the incubation on ice for a further 15 min.

   Note. Cells which have been chloramphenicol-treated in order to amplify the plasmid DNA content are often found to be fragile and are very easily 'over-lysed'. These cells should be treated with lysozyme and EDTA (steps 3 and 4) for 3–5 min only.

5 Add 4 ml 2% Triton X-100 in 50 mM Tris-HCl pH 8.0, 0.0625 M EDTA and mix to lyse the cells. The solution should become viscous and darken in colour as the cells lyse to release chromosomal DNA. If lysis fails to occur, incubate the cell suspension at 37°C for a few minutes.

6 Clear the lysate by centrifugation for 30 min at 18,000 rpm at 4°C. The cell debris should form a gelatinous but compact pellet. If the cells have been 'over-lysed' and a large fluffy white mass is produced which will not pellet, place the tube in a boiling waterbath for 5–10 min until the protein and chromosomal DNA are denatured. Chill the tube in ice and centrifuge a second time as before.

7 Decant the supernant liquid to a clean tube. This is the *cleared-lysate*, which can be used for dye-bouyant equilibrium centrifugation in caesium chloride/ethidium bromide gradients in order to recover the supercoiled plasmid DNA.

8 To 7·4 ml cleared lysate add 7·1 g CsCl and 0·5 ml 10 mg ml^{-1} ethidium bromide and mix. Ensure that all the caesium chloride has dissolved.

9 Transfer the solution to a 13 ml polyallomer centrifuge tube, fill the tube to the top with liquid paraffin, seal with a cap and centrifuge in a titanium fixed angle rotor (e.g. Beckman 50Ti) at 40,000 rpm at 19°C for 40 h.

10 Carefully remove the tube from the rotor and examine for bands using ultraviolet visualization if necessary. Two bands should be visible in the centre and in the lower half of the gradient. The lower band is supercoiled, covalently-closed circular plasmid DNA. The upper band is the dye-bouyant linear chromosomal and open-circular plasmid DNA. The pellicle at the top of the gradient is denatured protein and the pellet at the bottom RNA. The lower plasmid band can be collected by piercing the side of the tube just below the band with a 19 or 21 gauge needle and very gently removing the band with a 5 ml syringe. If the bands are close together it is sometimes advantageous to remove the upper band of chromosomal DNA first.

11 In order to avoid shearing the plasmid DNA, pour the syringe contents from the plunger end into a clean tube.

12 Remove the ethidium bromide by at least three successive extractions with an equal volume of isopropanol which has been equilibrated against caesium chloride saturated TE (10 mM Tris-HCl pH 8·0; 1 mM EDTA) buffer. Continue these extractions until all pink colouration has disappeared.

13 Dialyse the DNA solution against four changes of 2l TE buffer at 4°C in order to remove the caesium chloride.

14 Extract the DNA with an equal volume of redistilled phenol which has been equilibrated with T.E. buffer. Discard the lower phenol layer.

15 Remove remaining traces of phenol by three extractions with an equal volume of water-saturated diethyl ether and carefully blow compressed air into the tube to clear the ether vapours.

16 Finally concentrate the DNA by ethanol precipitation. Add 1/10 volume of 3 M sodium acetate (pH 6.0) and 2 vol of absolute ethanol. Mix and leave at −20°C overnight or at −70°C for at least 1 h. Pellet the DNA by centrifugation at 10,000 rpm for 30 min at 0°C. Decant off the supernatant, carefully rinse the pellet with 70% ethanol (−20°C) and dry the pellet *in vacuo.*

17 Resuspend the pellet of plasmid DNA in 0·5 ml TE buffer and determine the DNA concentration by measuring the absorbance at 260 nm of a diluted sample. For 1 cm path length, 50 μg ml^{-1} DNA solution has an absorbance at 260 nm of 1·0.

Method 2 (the NaOH method)

Chromosomal DNA in the cell lysate is selectively denatured with NaOH under conditions where covalently closed circular plasmid remains 'native'. On neutralization, chromosomal DNA forms an insoluble network. Most protein and ribosomal RNA is removed by precipitation with sodium dodecyl sulphate and high salt. This method is useful for screening DNA of single colonies.

Plasmid DNA is precipitated from the supernatant with EtOH and can be applied directly to agarose gels for analytical purposes. Small scale preparative amounts can also be prepared with minor modifications of the method.

1 Inoculate L-broth (1·5 ml) with a large loopful of cells. Use 6 ml capacity vials and shake at 37°C overnight.

2 Incubate 10–14 ml L-broth with 1·5 ml of pre-culture at 37°C. Shake until A_{600} = 10–12. Add chloramphenical 170 μg ml^{-1} using a stock solution with 34 mg ml^{-1} in EtOH. Incubate 16–18 h at 37°C with shaking.

3 Collect cells by centrifugation. Remove supernatant carefully with fine-tip aspirator. Suspend the pellet in 1·0 ml of a freshly-prepared lysis solution at 0°C for 30 min. The lysis solution contains lysozyme, 2 mg ml^{-1}; Tris-HCl, pH 8·0; 25 mM; CDTA or EDTA, pH 8·0, 10 mM; glucose, 50 mM (CTDA cyclohexane diamine tetra acetate from Aldrich Chem Co. is more soluble in alcohol than EDTA).

4 Add 2·0 ml NaOH/SDS solution (0·2 M NaOH; 1% SDS) at 0°C for 5 min. Transfer with Pasteur pipette to 15 ml Corex centrifuge tube. Add 1·5 ml 3 M Na acetate. Adjust pH to 4·8. Mix gently by inversion (a clot of precipitated chromosomal DNA should be visible), keep at 0°C for 60 min. Centrifuge 10,000 rpm at 20°C for 10 min.

5 Remove 4·0 ml of clear supernatant to fresh 15 ml Corex tube. Add 8·0 ml of cold ethanol. Stand for 15 min at −20°C.

Centrifuge at 10,000 rpm at 0°C for 5 min. Discard supernatant.

6 Suspend pellet in 2·0 ml of 0·1 M Na acetate; 1 mM EDTA; 0·1% SDS; 40 mM Tris, pH 8.0. Add 2·0 ml phenol/chloroform (100 g phenol crystals dissolved in 100 ml CHCl$_3$). Vortex gently for 3–4 min at room temperature. Centrifuge 5,000 rpm at 20°C for 2 min.

7 Carefully remove aqueous phase I. Re-extract interface once with 1·0 ml of acetate/EDTA/SDS/Tris as before. Add 2 vol EtOH to combined aqueous phases, keep at −20°C for 15 min. Centrifuge 10,000 rpm, 0°C, 5 min. Discard supernatant. Resuspend pellet in 200 μl 10 mM Tris; 1 mM EDTA. Extract once with neutralized phenol, ether extract, then precipitate with ethanol. Estimated recovery of pBR322 DNA, about 100–200 μg 10 ml^{-1} culture.

Method 3

Considerable difficulty is often encountered in recovering large non-amplifiable plasmids by many standard procedures, including the two indicated above. This may be due to shearing, to nicking (of relaxation complexes) during preparation or simply to non-specific association of the plasmid with the chromosomal DNA fraction which is removed at an early stage. One method which we have used for large plasmids with some success is the following one based upon that described by Hansen and Olsen.

1 Harvest an overnight culture at 250 ml of the host strain in Luria Broth by spinning at 7,000 rpm for 10 min. Suspend the pellet in 6 ml sucrose (25% in 0·05 M Tris-HCl, pH 8·0), and lyse in the following way.

2 Add 1 ml lysozyme (5 ml ml^{-1} in 0·25 M Tris, pH 8·0). Mix by inversion and keep at room temperature for 5 min; add 2·5 ml EDTA (0·25 M, pH 8·0), mix and incubate for a further 5 min at room temperature. Finally, add 2·5 ml 20% SDS in TE buffer; mix to complete lysis, incubating briefly at 55°C if necessary.

3 Add 0·75 ml 3 M NaOH to denature DNA. Mix gently for 3 min. Add 6.0 ml 2 M Tris, pH 8.0 and mix thoroughly but without shearing. This treatment should bring the pH back to about 8·5. Add 3 ml 20% SDS followed by 6 ml ice cold 5 M NaCl. Mix thoroughly immediately. Leave overnight at 4°C or 5–6 h on ice.

4 Spin down the flocculent precipitate at 20,000 rpm for 30 min. To the supernatant add ⅓ vol of 42% polythylene glycol 6,000 (in 0·1 M phosphate buffer, pH 7·0), mix and incubate over night at 4°C.

5 Spin down the precipitated DNA (7,000 rpm for 6 min). Resuspend by gentle shaking in 8 ml of TE buffer and finally band in a CsCl gradient in the presence of ethidium bromide as above. The method can yield up to 50 µg of plasmid DNA.

Reference

Hansen, J.B. & Olsen, R.H. (1978) Isolation of large bacterial plasmids and characterisation of the P2 incompatibility group plasmids, pMG1 and pMG5. *J. Bacteriol.*, **135**, 227.

Section B
Analysis of DNA and RNA

Alec Jeffreys
Jayne Mathews
Peter Williams

Introduction

Techniques for the fine-structure analysis of nucleic acids such as cloned fragments of DNA are a vital part of recombinant DNA research. Such techniques include the analysis of DNA molecules by cleavage with restriction endonucleases, nucleic acid hybridization, sequencing and electron microscopy.

The objectives of these practicals are:

1 To map restriction endonuclease cleavage sites in plasmid DNA by: (1) analysis of single and double digest products; (2) Smith-Birnstiel mapping.
2 To detect and map mammalian genes by Southern blot hybridization.
3 To use spot hybridizations to estimate sequence homology.
4 To detect mRNAs by Northern blot analysis.

The experiments in this Section can be accommodated in a 4-day period if they are carried out in parallel. Experiments B.2, B.3 and B.4 have a number of features in common which can be carried through together.

Experiment B.1 Construction of plasmid restriction maps

Background

Plasmid pBR322 is one of the most widely used vectors in cloning work. It is a multicopy plasmid (based on the ColEl replicon) carrying ampicillin- and tetracycline-resistance genes (Fig. B.1) and was derived as outlined in Fig. B.2. It has a single cleavage site for a number of commonly used restriction enzymes and its entire nucleotide sequence (4·3 kb) is known (Sutcliffe 1978). Plasmid pAT153 is a derivative of pBR322. It was constructed by deleting a region of the parent plasmid (Fig. B.2) which has the effect of raising copy number (Twigg & Sherratt 1980). It retains the two resistance genes and all but one of the important single restriction sites. Thus, as well as the technical convenience of reduced size (3·6 kb), pAT153 has

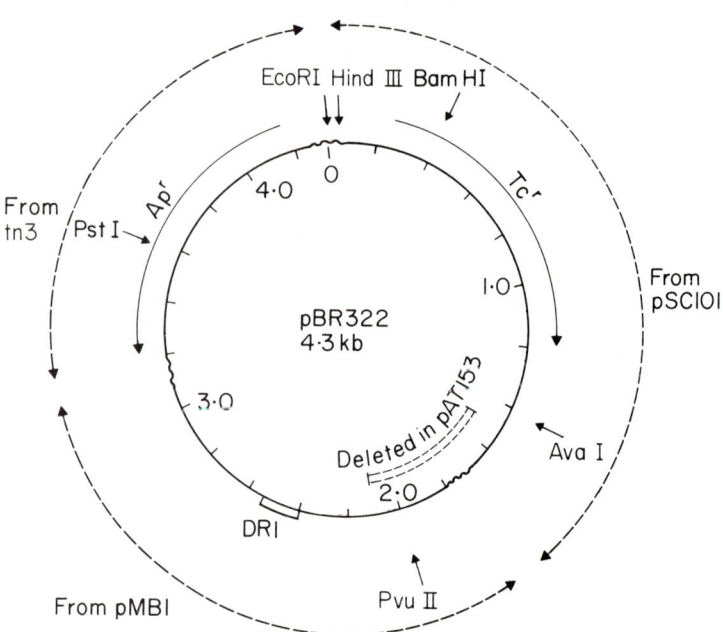

Fig. B.1 Biological map of the multi copy plasmid pBR322. Approximate positions of the junctions of fragments from pSC101 (Tc^r), tn3 and pBM1 (contributing the origin of replication) are shown (~~). Transposon tn3 appears to have an intact left-hand end near pMB1 but has probably lost the right-hand end (Sutcliffe 1979). The origin of replication region is almost identical with that of ColEl suggesting that pMB1 is another naturally occurring El-plasmid. Origins of transcription from the *bla* Ap^r and Tc^r promoters are indicated as are the positions of unique restriction enzyme sites. Region = = = = is deleted in pAT153.

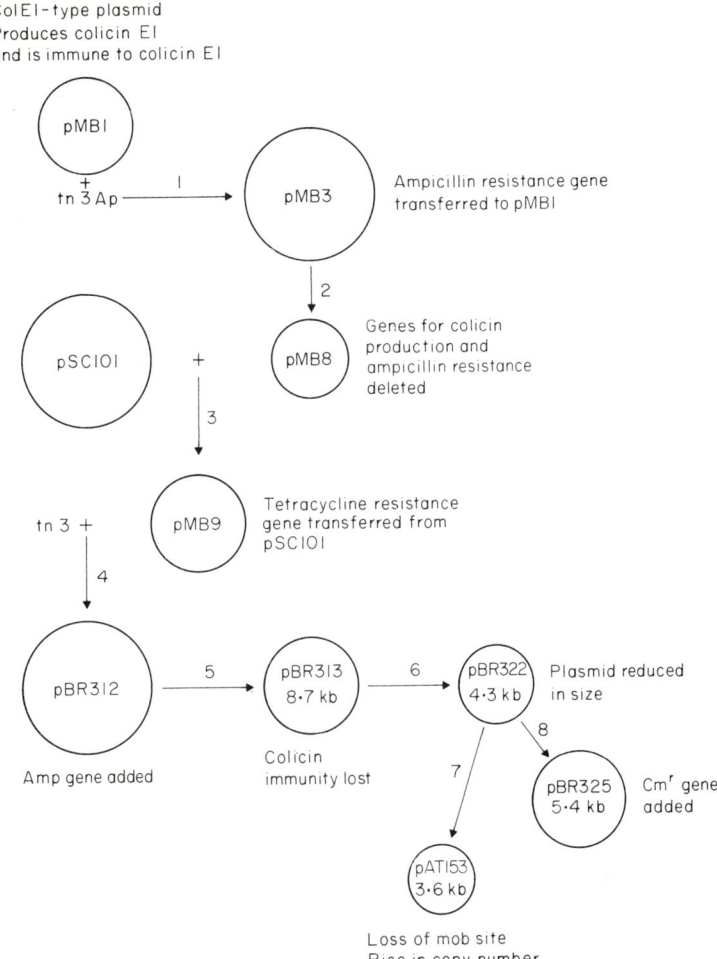

Fig. B.2 Plasmid pMB1 is a naturally occurring ColE1-type plasmid. (1) pMB3 was derived by transposition of tn3 (Ap) into pMB1 such that colicin production and immunity were unaffected. (2) Following EcoRI digestion and ligation, a single fragment of pMB3 was recovered as plasmid pMB8, which specifies colicin immunity but lacks determinants for colicin production and ampicillin resistance. (3) A fragment of pSC101 carrying the tetracycline resistance determinant was ligated into the *Eco*RI site of pMB8 to generate pMB9, and (4) pBR312 was obtained by transposition of tn3 into a nonessential region of pMB9. (5) Plasmid pBR312 was reduced in size by partial *Eco*RI digestion and ligation, giving pBR313, with loss of the colicin immunity determinant and (6) further reduced to less than half size to give pBR322. (7) Deletion of two *Hae*II fragments of pBR322 yielded pAT153 which cannot be mobilized due to absence of the mob site, and (8) insertion of a part of tn9, which confers resistance to chloramphenicol, generated pBR325.

the advantage of having up to three times as many copies per cell as pBR322.

In this section of the course a restriction map of plasmid pAT153 will be constructed.

Part a is a conventional mapping project involving analysis of sizes of fragments generated by single and double digestion with five enzymes each of which cleave the plasmid only once or twice.

Part b uses the Smith and Birnsteil technique (Smith & Birnsteil 1978) to map the cleavage sites for enzymes which each cut pAT153 many times as described in detail by Boseley *et al.* (1980). Plasmid DNA will be linearized by cleavage with *Bam*Hl and labelled by filling in the recessed 3′ ends generated using the Klenow fragment of *E. coli* DNA polymerase I in the presence of α-^{32}P-dCTP and unlabelled dGTP, dATP and dTTP. End-labelled DNA will then be cut asymmetrically with *EcoRI* and subjected to partial digestion with *Dde1*, *Hin*fI and *Alu*I. Subsequent analysis will involve electrophoresis of the partial digestion products in an agarose gel, followed by autoradiography. This technique provides a rapid method for the fine structure mapping of multiple enzyme cleavage sites in a DNA molecule; by contrast, determination of such maps by conventional methods of double, triple or more complex digests is often very difficult.

For successful exploitation of the Smith and Birnsteil technique a restriction enzyme which cleaves the linearized DNA close to one end should be used. Thus a preliminary restriction map [part (a)] is required. In this course, however, part (b) is started before part (a) in order to complete the experiments in a 4-day period.

References

Boselly, P.G, Moss, T. & Birnsteil, M.L. (1980) 5′ labelling and poly (dA) tailing. *Meth. Enzym.*, **65**, 478.

Smith, H.O. & Birnsteil, M.L. (1976) A simple method for DNA restriction site mapping. *Nucl. Acid. Res.*, **3**, 2387.

Sutcliffe, J.G. (1978) pBR322 restriction map marked from the DNA sequence: accurate DNA size markers up to 4361 nucleotide pairs long. *Nucl. Acid. Res.*, **5**, 2721.

Sutcliffe, J.G. (1979) Complete nucleotide sequence of the *Escherichia coli* plasmid pBR322. *Cold Spring Harbor Symp. Quant. Biol.*, **43**, 77.

Twigg, A.G. & Sherratt, D.J. (1980) Trans complementable copy-number mutants of plasmid ColE1. *Nature*, **283**, 216.

(a) Mapping cleavage sites for *Eco*RI, *Hin*dIII, *Bam*HI, *Pst*I and *Hin*cII in plasmid pAT153

One microgram amounts of plasmid DNA will be digested with restriction enzymes singly and in pairs. In cases where enzymes require similar reaction conditions it is possible to perform both reactions of a double digest simultaneously. When enzymes require different buffers it is necessary either to modify the reaction mixture after one of the digestions of a double digestion or to precipitate the DNA between reactions. Since the buffer conditions required for *Eco*RI are different from those of the other enzymes used in this part of the experiment, stocks of pAT153 DNA already cleaved with *Eco*RI are used as well as uncut plasmid, in order to save time.

The following restriction digests should be made:

Group 1	Group 2	Group 3
*Hin*dIII	*Eco*RI	*Pst*I
*Eco*RI + *Hin*dIII	*Bam*HI	*Hinc*II
*Eco*RI + *Pst*I	*Eco*RI + *Bam*HI	*Eco*RI + *Hinc*II
*Hin*dIII + BamHI	*Bam*HI + PstI	*Pst*I + *Hinc*II
*Hin*dIII + PstI	*Bam*HI + *Hinc*II	*Hin*dIII + *Hinc*II

Materials needed

pAT153 DNA (200 µg ml^{-1})
*Eco*RI-digested pAT153 DNA (200 µl ml^{-1})
5 × *Eco*RI salts mix (250 mM NaCl; 0·5 M Tris-HCl, pH 7·5; 50 mM MgCl$_2$)
5 × salts mix for *Hin*dIII, *Bam*HI, *Pst*I and *Hinc*II (250 mM NaCl; 30 mM Tris-HCl, pH 7·5; 30 mM MgCl$_2$; 30 mM 2-mercaptoethanol)
Enzymes (1 unit µl^{-1})
Sterile water
Loading buffer for agarose [10 ml Analar glycerol; 8 ml 0·5 mM EDTA, pH 7·4; 2 ml water; small amount (< 1 mg) of bromophenol blue]
Agarose
Ethidium bromide (5 mg ml^{-1})
Electrophoresis buffer (13 mM Tris-HCl, 0·33 mM EDTA adjusted to pH 7·7 with glacial acetic acid)
Molecular weight markers

Procedure

1 Set up the following 15 µl reaction mixtures in Eppendorf tubes: 5µl DNA, 3 µl 5 × approximate salts mix and 1 µl each enzyme. Make up to 15 µl with water.
2 Centrifuge briefly to collect the whole reaction mixture in the bottom of the tube.
3 Incubate at 37°C for 1 h.
4 Add 5 µl loading buffer and run on a 7.0% agarose gel containing ethidium bromide (0.5 µg/ml). Also run molecular weight markers.
5 Photograph gel by transmitted short-wave ultra-violet light.
6 Measure relative mobilities of marker DNA and restriction fragments; determine fragment sizes and construct a map of

EcoRI, HindIII, BamHI, PstI and HincII cleavage sites in pAT153.

(b) Mapping DdeI, HinfI and AluI cleavage sites in pAT153 using end-labelled linear DNA and partial enzyme digestion

Preparation of linear pAT153 DNA

Materials needed

pAT153 DNA (200 µg ml^{-1})
5×BamHI salts mix (250 mM NaCl; 30 mM Tris-HCl, pH 7·4; 30 mM MgCl$_2$; 30 mM 2-mercaptoethanol)
BamHI (2 units µl^{-1})
Loading buffer
Phenol mix (phenol:chloroform:isoamyl alcohol:8-hydroxyquinoline, 100:100:4:0·1).
TE buffer (10 mM Tris-HCl, pH 7·5; 1 mM EDTA)
Chloroform:isoamyl alcohol (24:1)
2 M sodium acetate, pH 5·6)
Ethanol.

Procedure

1   Set up the following reaction in an Eppendorf tube: 25 µl pAT153, 7µl 5×BamHI buffer mix and 3 µl BamHI enzyme.
2   Mix thoroughly and centrifuge briefly to collect the whole reaction mixture to the bottom of the tube.
3   Incubate for 1 h at 37°C.
4   Remove 3·5 µl (0·5 µg) from reaction mix, add 1 µl of loading buffer and run on a 0·5% mini-agarose gel (see Experiment A.11). Also run 2·5 µl of uncut DNA. Store the remainder of the reaction mixture on ice and, if the digestion is subsequently found to be complete, phenol extract the DNA to remove endonuclease as follows:
5   Add an equal volume of phenol mix to the reaction mixture, mix thoroughly and centrifuge for 2 min in the microcentrifuge.
6   Remove upper (aqueous) phase to a fresh Eppendorf tube.
7   Re-extract phenol layer by adding 50 µl of TE. Mix well and centrifuge for 2 min.
8   Remove aqueous phase and pool with the first.
9   Extract residual phenol from pooled aqueous fractions by adding an equal volume of chloroform:isoamyl alcohol (24:1). Mix well and centrifuge for 2 min.

10 Remove upper (aqueous) phase to a fresh Eppendorf tube.
11 Add approximately 1/10 vol of 2 M sodium acetate, pH 5·6, and 2 vol of ethanol. Mix well and chill in an ethanol/dry-ice bath for 5 min to precipitate DNA.
12 Centrifuge for 5 min.
13 Carefully remove supernatant with a Pasteur pipette and vacuum dry pellet.
14 Resuspend pellet in 25 µl distilled water and store at $-20°C$.

End labelling with Klenow enzyme

Materials needed

$10 \times$ nick mix (0·5 M Tris-HCl, pH 7·8; 50 mM $MgCl_2$; 0·1 M 2-mercaptoethanol)
Unlabelled triphosphates (dGTP, dATP and dTTP, 2 mM each)
α-^{32}P-dCTP (10 mCi ml^{-1} in aqueous solution, 2,000–3,000 Ci mmol^{-1})
Klenow enzyme (10 units µl^{-1})
Phenol mix
TE buffer
Chloroform : isoamyl alcohol (24 : 1)
tRNA (5 mg ml^{-1} E. coli tRNA, Sigma)
2 M sodium acetate, pH 5·6
Ethanol
70% ethanol

Procedure

1 Thaw out linearized DNA and set up the following reaction in an Eppendorf tube: 25 µl DNA in water, 3·5 µl $10\times$ nick mix, 2·5 µl unlabelled dGTP, dATP and dTTP (2 mM each), 2 µl α-^{32}P-dCTP and 2 µl Klenow enzyme.
2 Incubate for 1 h at room temperature.
3 Phenol extract reaction mixture as described (add 50 µl phenol mix, re-extract the phenol layer with 50 µl TE, and extract residual phenol from pooled aqueous fractions with 100 µl chloroform : isoamyl alcohol).
4 Add 10 µl tRNA (5 mg ml^{-1} E. coli tRNA) as carrier.
5 Ethanol precipitate DNA as described above (add sodium acetate to 0·2 M and 2 vol of ethanol, chill and centrifuge). Resuspend pellet (drying is not necessary) in 100 µl 70% ethanol.
6 Chill and centrifuge as above. Resuspend pellet (without drying) in 100 µl 70% ethanol.

7 Chill and centrifuge again (in order to ensure removal of all unincorporated α-^{32}P-dCTP). Vacuum dry pellet and resuspend in 50 µl water.

8 Cerenkov count the solution, and save a small sample containing about 100,000 cpm.

*Eco*RI digestion

Materials needed

$5 \times Eco$RI salts
EcoRI (10 units µl^{-1})
2 M sodium acetate, pH 5·6
Ethanol
Loading buffer
Labelled molecular weight markers
X-ray film

Procedure

1 Set up the following reaction in an Eppendorf tube: 50 µl DNA sample, 14 µl $5 \times Eco$RI buffer and 5 µl *Eco*RI enzyme.
2 Incubate for 1 h at 37°C.
3 Ethanol precipitate DNA, vacuum dry pellet and Cerenkov count.
4 Redissolve pellet in water at 50,000 cpm µl^{-1}.
5 Remove a 2 µl sample (freeze the remainder at −20°C), add 1 µl loading buffer and run on a 0·5% mini-agarose gel, along with the sample you saved from Klenow reaction. Also run the radioactive markers.
6 Transfer the gel to a large glass plate and dry down using a hair-dryer.
7 Expose the dried gel to X-ray film for 1 h at room temperature and develop the autoradiograph as demonstrated. This gel allows you to check the efficiency of incorporation of label in the Klenow reaction, and to monitor the *Eco*RI digestion.

Partial digestion and agarose gel electrophoresis

Materials needed

$5 \times Dde$I salt mix (0·5 M NaCl; 0·5 M Tris-HCl, pH 7·5; 25 mM MgCl$_2$; 30 mM 2-mercaptoethanol)
$5 \times Hinf$I salt mix (0·5 M NaCl; 30 mM Tris-HCl, pH 7·5; 30 mM MgCl$_2$; 30 mM 2-mercaptoethanol)

5×*Alu*I salt mix (250 mM NaCl; 250 mM Tris-HCl, pH 8·0; 25 mM $MgCl_2$; 5 mM dithiothreitol)
Restriction enzymes (10 units μl^{-1})
0·1 M EDTA
Loading buffer
Labelled molecular weight markers
X-ray film

Procedure

Use the following restriction enzymes: Group 1, *Dde*I; Group 2, *Hin*fI; Group 3, *Alu*I.

1   Thaw sample and set up the following reaction mixtures in an Eppendorf tube on ice: 10 µl DNA, 2·5 µl 5×appropriate salt mix and 1 µl enzyme.
2   Mix thoroughly and centrifuge briefly to collect the total reaction mixture at the bottom of the tube.
3   Immediately remove the equivalent of approximately 10,000 cpm to a fresh tube and add 7 µl 0·1 M EDTA and 2 µl loading buffer. Store on ice (undigested sample).
4   Transfer reaction tube to a 37°C water bath.
5   At 1 min, 2 min, 4 min and 8 min remove the equivalent of approximately 10,000 cpm to a single fresh tube (i.e. pool the samples) containing 4 µl 0·1 M EDTA and 2 µl loading buffer. Store on ice (partial digests).
6   Take a final sample of approximately 10,000 cpm after 60 min. Add 7 µl M EDTA and 2 µl loading buffer. Store on ice (complete digestion).
7   Run samples, together with the radioactive molecular weight markers provided, on a 0.5% agarose gel at 100 V for about 3 h.
8   Dry down gel using hair-dryers.
9   Autoradiograph overnight at −70°C (see Appendix C.III).

Analysis and interpretation

1   Develop autoradiograph.
2   Measure relative mobilities of marker DNA and samples, and determine sizes of partial digestion products.
3   Add *Dde*I, *Hin*fI and *Alu*I cleavage sites to your restriction map of pAT153.

Experiment B.2 Detection of mammalian β-globin genes in total genomic DNA by filter hybridization with cloned β-globin cDNA

Human and rabbit DNAs will be digested with various restriction endonucleases, denatured and electrophoresed through an agarose slab gel. The DNA fragments separated by this technique will be transferred to a nitrocellulose filter by blotting (Southern 1980). Rabbit adult β-globin cDNA cloned into the plasmid pMB9 (Maniatis *et al.* 1976) will be labelled with ^{32}P *in vitro* by nick translation and used as a hybridization probe for β-globin genes. The probe is denatured and hybridized to the filter containing mammalian DNAs. After hybridization, unbound labelled probe is washed from the filters, and labelled DNA bands, containing β-globin genes, are detected by autoradiography.

(a) Preparation of mammalian DNA

Various methods have been published, not all of which give reliable substrates for restriction endonucleases, nick translation, etc. The method given in Appendix D.II gives reliable high molecular weight preparations and is suitable for isolating DNA from any animal source (Jeffreys 1979).

(b) Restriction endonuclease digestion of mammalian DNA

Sufficient mammalian DNA has to be restricted to permit the detection of a single copy gene in Southern blot hybridizations. In general, 10 μg DNA are digested for each gel track. This amount of DNA contains about 3 pg of a given 1 kb single copy gene, and this quantity of sequence can be readily detected by hybridization.

Procedure

1 Mix 100 μl 100 μg ml^{-1} mammalian DNA, 25 μl 5× restriction endonuclease buffer (according to manufacturer's specification) and 10 units restriction endonuclease. Digest at the recommended temperature for 1 h.

2 Remove 10 µl of the digest and electrophorese through a 0·4% agarose gel (Appendix B.II). View the gel under an ultraviolet light. A typical digest profile is shown in Fig. B.3 and, with experience, completion of the digestion can be judged from such patterns. Alternatively, λ DNA can be included to monitor the digestion.

Recovery of digested DNA

3 Add 1/20 vol 0·5 M EDTA, pH 8·0 and 1 vol phenol/chloroform, mix, spin in an Eppendorf centrifuge for 1 min.
4 Re-extract the phenol phase with ½ vol H_2O.
5 Pool the aqueous phases, add 1/10 vol 2 M NaAc, pH 5·6 and 2·5 vol ethanol. Chill in an ethanol/dry-ice bath for 2 min and pellet the DNA in an Eppendorf centrifuge for 5 min.
6 Rinse the pellet with 70% ethanol, vacuum dry and redissolve the DNA in 10 µl H_2O.

(c) Electrophoresis and transfer of DNA to filters

Materials needed

1 mg ml^{-1} rabbit liver DNA×*Eco*RI
1 mg ml^{-1} rabbit liver DNA×*Eco*RI×*Pst*I
1 mg ml^{-1} rabbit liver DNA×*Pst*I
1 mg ml^{-1} human blood DNA×*Pst*I: individual 1
1 mg ml^{-1} human blood DNA×*Pst*I: individual 2
Marker DNA: 0·4 µg ml^{-1} λ DNA×*Hind*III in 1 mg ml^{-1} rabbit DNA×*Eco*RI (λ marker fragment sizes: 23·5, 9·6, 6·8, 4·4, 2·3, 1·9, 0·58 kb)
Gel-loading slurry prepared as described in Appendix B.II.

Procedure

Prepare a slab gel with 0·3% agarose as described in Appendix B.II.

Electrophoresis of DNA digests

Mix 10 µl of each DNA digest or marker DNA with 10 µl agarose beads and 2 µl 1·5 M NaOH, 0·1 M EDTA.
Load onto the gel according to the loading pattern in Table B.1 and leave for 5–10 min before running.

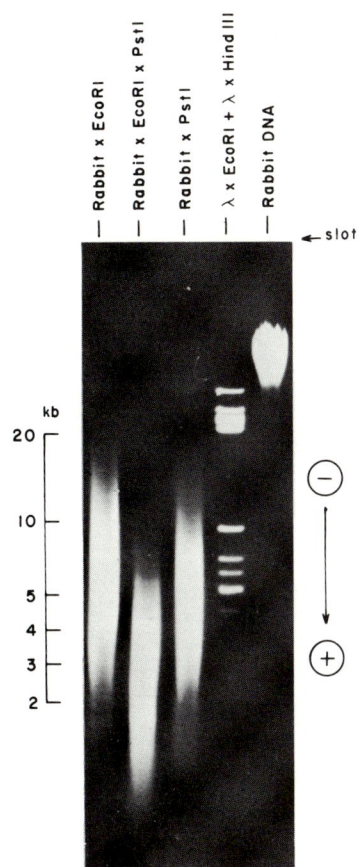

Fig. B.3 Electrophoretic profiles of rabbit DNA digested with various restriction endonucleases. DNA samples (1 µg per slot) were electrophoresed in a 0·4% agarose gel containing ethidium bromide, and DNA was visualized under an ultraviolet transilluminator. Note that before digestion, almost all of the rabbit DNA is high molecular weight (> 50 kb).

Table B.1. DNA loading pattern

1 Marker DNA
2 Rabbit×*Eco*RI
3 Rabbit×*Eco*RI×*Pst*I
4 Rabbit×*Pst*I
5 Individual 1×*Pst*I
6 Individual 2×*Pst*I
7 Rabbit×*Eco*RI
8 Rabbit×*Eco*RI×*Pst*I
9 Rabbit×*Pst*I
10 Individual 1×*Pst*I
11 Marker DNA

Electrophoresis: run the gel at 150 V (8 V cm^{-1}) for about 2 h until the bromophenol blue has migrated about 8 cm.

Blotting the DNA agarose gel (see Southern 1980)

Note. Since the DNA was denatured before electrophoresis it can be transferred directly to a nitrocellulose filter after the run.

1 Place the gel in a photographic tray.
2 View under ultraviolet light to check for a uniform run.
3 Wash the gel for 20 min in $20 \times$ SSC
($1 \times$ SSC = 0.15 M NaCl, 15 mM Na citrate, pH 7.0).
4 Set up the transfer apparatus as shown in Fig. B.4. Assemble as follows:

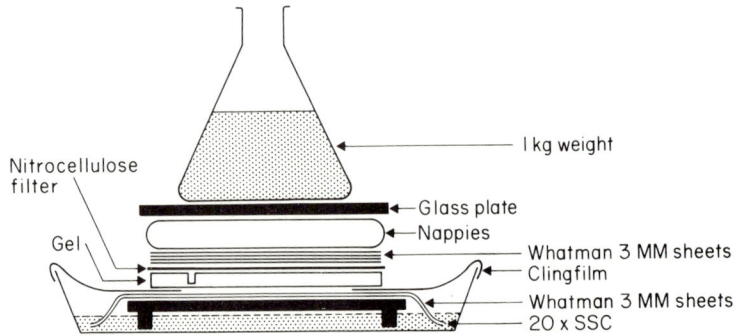

Fig B.4 Southern blotting apparatus. Note that the lower part of the apparatus can be reused indefinitely and can be stored without a gel provided that the 'window' in the Clingfilm is kept covered.

(a) Place the glass plate with supports in a tray flooded with $20 \times$ SSC.
(b) Cover with two sheets of Whatman 3 MM paper soaked in $20 \times$ SSC; these sheets dip into the tray and act as wicks.
(c) Cover paper and tray with Clingfilm; fix the Clingfilm to the sides of the tray with tape.
(d) Cut a hole in the Clingfilm slightly smaller than the slab gel.
(e) Soak the exposed paper with $20 \times$ SSC and place the gel in position, avoiding air bubbles.
(f) Cut a Sartorius nitrocellulose filter (0.45 μ pore size) to the correct size and soak in $3 \times$ SSC. (Note. Use gloves when handling filters.)
(g) Place the filter on the gel, avoiding air bubbles.
(h) Soak a 20×20 cm Whatman 3 MM filter in $3 \times$ SSC, place on the nitrocellulose filter (no bubbles).
(i) Remove surplus liquid from the gel.

(j) Place three dry 3 MM filters on top, followed by a wad of disposable nappies, a glass plate and a weight.
(k) Leave to transfer at 4°C overnight.

Baking the nitrocellulose filter

1 Remove the nitrocellulose filter from the gel, marking the uppermost side with pencil.
2 Rinse for 3 min in 3×SSC.
3 Blot dry on Whatman 3 MM paper.
4 Bake in an 80°C oven for 4 h to fix the DNA to the filter.

(d) Preparation of ^{32}P-labelled probe DNA by nick translation

Background

Nick translation provides a rapid method for the *in vitro* labelling of DNA with ^{32}P (Rigby *et al.* 1977). DNA is nicked with DNase I and the exonuclease activity of *E.coli* DNA polymerase I converts these nicks to single-stranded regions that are filled in by polymerase I. In the presence of α-^{32}P-dNTPs, radioactive DNA is synthesized. After nick translation, the reaction is terminated by phenol extraction and the labelled DNA recovered by ethanol precipitation in the presence of high molecular weight carrier DNA.

Recombinant DNAs are frequently used as hybridization probes. In this experiment, rabbit adult β-globin cDNA cloned into the plasmid pMB9 (Maniatis *et al.* 1976) is used to detect β-globin genes. While the entire recombinant, termed PβG1, can be used as a probe, most (90%) of the label is in plasmid DNA, not cDNA. PβG1 DNA is therefore cleaved with restriction endonuclease *Hha*I and a 1·5 kb fragment containing the entire 0·6 kb cDNA insert recovered by preparative gel electrophoresis (Smith 1980; see Appendix B.III). This fragment is used as a substrate for nick translation.

Materials needed

10×nick mix (0·5 M Tris-HCl, pH 7·8, 50 mM MgCl$_2$, 0·1 M 2-mercapto-ethanol)
50 μM dATP
50 μM dGTP
50 μM dTTP
DNse I (100 μg ml^{-1} in 10 mM Tris-HCl, pH 7·5; Worthington).
10 mM Tris-HCl, pH 7·5
E. coli DNA polymerase I (4 u μl^{-1}; Boehringer grade I).

α-^{32}P-dCTP (10 mCi ml^{-1} in aqueous solution; 2000–3000 Ci/mmol, Amersham)
Quench mix (2% SDS, 50 mM EDTA, 10 mM Tris-HCl, pH 7·5)
Phenol/chloroform mix (phenol:chloroform:isoamyl alcohol:8-hydroxyquinoline–100:100:4:0.1 (w:v:v:w); store under 10 mM Tris-HCl, pH 7·5)
2 M Na acetate, pH 5·6
Salmon DNA (high molecular weight, 1 mg ml^{-1} in water, Sigma)
70% ethanol
Ethanol
Substrates:
15 μg ml^{-1} 1·5 kb cDNA fragment isolated from PβG1 DNA digested with *Hha*I.
15 μg ml^{-1} λ phage DNA (to detect marker DNA bands)

Procedure

1   Mix on ice in an Eppendorf tube:
    5 μl (75 ng) DNA
    2·5 μl 10×nick mix
    2 μl 50 μM dATP
    2 μl 50 μM dGTP
    2 μl 50 μM dTTP
    8·5 μl H$_2$O
    1 μl 8 ng ml^{-1} DNAase I (Worthington)—freshly diluted into 10 mM Tris-HCl, pH 7·5, from a 100 μg ml^{-1} stock solution
    1 μl *E.coli* DNA polymerase I
    2 μl α-^{32}P-dCTP
2   Incubate 15°C for 45 min.

Recovery of labelled DNA

3   Add 25 μl quench mix.
4   Add 50 μl phenol/chloroform.
5   Mix, centrifuge at full speed for 1 min in a bench-top centrifuge, with tube within a 30 ml Corex tube to prevent ^{32}P leakage.
6   Remove the aqueous layer.
7   Re-extract the phenol layer with 50 μl 10 mM Tris-HCl, pH 7·5, with a spin.
8   Pool the aqueous layers in a small glass test-tube.
9   Add 20 μl 2 M Na acetate, pH 5·6.
10 Add 100 μl 1 mg ml^{-1} salmon DNA, mix.
11 Add 500 μl ethanol, mix; the DNA should appear as a fibrous clot.

12 Remove liquid with a siliconized pasteur pipette, rinse the DNA clot with 70% ethanol.
13 Dissolve DNA in 200 µl 10 mM Tris-HCl, pH 7·5. Then add 20 µl 2 M Na acetate and 440 µl ethanol.
14 Mix, rinse DNA with 70% ethanol, dissolve in 500 µl 10 mM Tris-HCl.
15 Estimate incorporation by Cerenkov counting. At least 10^6 cpm ^{32}P should be incorporated into DNA.

(e) Hybridization of the DNA filter with ^{32}P-labelled DNA

Background

In order to detect the picogram amounts of β-globin DNA fragments in the filter-bound mammalian digests by hybridization with ^{32}P-labelled β-globin cDNA, the greatest care has to be taken to prevent aspecific binding of probe to the filters under the hybridization conditions we use (in 1×SSC at 65°C overnight; see Jeffreys *et al.* 1980). The filters are initially put through the ficoll-bovine serum albumin-polyvinylpyrrolidone treatment of Denhardt (1966) to block further aspecific binding of DNA to filters. Filters are then exposed to competitor salmon DNA and treated with dextran sulphate before addition of heat-denatured ^{32}P-DNA.

The dextran sulphate greatly increases the filter hybridization kinetics (Wahl *et al.* 1979, Jeffreys *et al.* 1980). Not all dextran sulphates are effective, and the best so far found is Sigma's molecular weight 500,000 dextran sulphate. Under these conditions, the concentation of probe should be no more

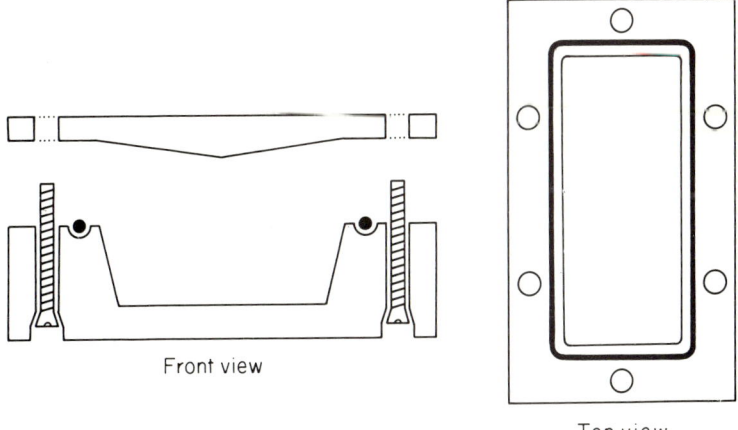

Fig. B.5. The hybridization box. The box and lid are made from heat-annealed perspex. The chamber dimensions are 10×4 cm, or larger if required, and 2·5 cm deep. The lid has a condensation drip point.

than 10 ng ml^{-1} ^{32}P-DNA; any more, and severe background labelling can occur.

The prehybridization and hybridization steps are performed in flat-bottom perspex chambers (see Fig. B.5). Circular chambers of a similar design can also be used for screening recombinant bacteriophage plaques (Benton Davis screening) or bacterial colonies harbouring recombinant plasmids (Grunstein-Hogness screening), using precisely the same hybridization protocol.

All hybridization solutions are de-gassed under vacuum prior to use to prevent aspecific patches of labelling which result from air bubbles appearing between filters during the hybridization (Jeffreys & Flavell 1977).

Materials needed

3×SSC

Denhardt's solution (0·2% ficoll-400; 0·2% bovine serum albumin; 0·2% polyvinylpyrrolidone in 1×SSC)

Solution A: Denhardt's solution supplemented with 0·1% SDS and 50 µg ml^{-1} low molecular weight salmon DNA

Solution B: Solution A supplemented with dextran sulphate (Sigma molecular weight 500,000; 9 g dextran sulphate per 100 ml solution A)

These solutions should be stored at −20°C

Low molecular weight competitor DNA: Salmon DNA (Sigma) is dissolved in water to ~ 5 mg ml^{-1}. NaOH and EDTA are added to 0·3 M and 20 mM respectively and the denatured DNA solution heated in a boiling water bath for 20 min. The solution is neutralized by addition of HCl, phenol extracted and the sheared single-stranded DNA collected by ethanol precipitation and dissolved in water.

Procedure

1 View the baked DNA filter under ultraviolet light; the DNA tracks will be visible by the fluorescence of bound ethidium bromide.

2 Use a scalpel to cut the filter between DNA tracks to give strips 3–3·5 cm wide by 9 cm long. These strips will fit into the hybridization chamber shown above. Mark the strips with pencil to aid identification.

3 Cut two blank 3·5×9 cm nitrocellulose filters.

4 Place a blank filter, then each baked strip (for the same probe) then a final covering blank filter into 30 ml 3×SSC in a hybridisation chamber.

5 Incubate at 65°C for 10 min in a slowly rocking water bath.

Note. All following solutions must be de-gassed under vacuum before use.

6 Transfer all strips sequentially to a box containing 30 ml Denhardt's solution.

7 Incubate at 65°C for 40 min.

8 Transfer strips sequentially to 20 ml solution A, incubate at 65°C for 40 min.

9 Transfer strips sequentially to 20 ml of Solution B. Mix the strips and solution thoroughly. Incubate at 65°C for 40 min.

10 Heat the ^{32}P-labelled DNA (0·5 ml in a glass test-tube) in a boiling waterbath for 5 min to denature the probe. Transfer to a box containing 10 ml solution B. Mix in the probe thoroughly.

11 Transfer the strips (flicking to remove excess liquid) into the hybridization solution, making sure each strip is thoroughly mixed with the solution. Incubate at 65°C overnight in a slowly rocking waterbath.

(f) Washing hybridized DNA filters

Background

After hybridization, filters are washed extensively under hybridization conditions (minus dextran sulphate) to remove non-hybridized probe. The hybridization and washing are carried out in 1×SSC at 65°C, about 18°C below the T_m of a perfectly matched DNA hybrid. Since the T_m of a DNA duplex is depressed by approximately 1°C for each % mismatch of sequence, only hybrids at least 82% homologous to the probe should form. More diverged hybrids can only be detected by hybridizing at a lower stringency (lower temperature or higher concentration of salt). If too low a stringency is used, many totally aspecific hybrids will form and will mask any authentic hybrids. The threshold for this problem is typically at about 60°C in 3×SSC.

Under the hybridization conditions used in this experiment, hybrids up to 18% divergent will form on the filter. The relatively mismatched hybrids may be selectively removed by washing under more stringent conditions (lower salt concentrations). Perfectly matched hybrids will only begin to melt from the filter at < 0·1×SSC at 65°C.

Materials needed

Solution A (see Experiment B.2, part e)
SSC wash mixes: 0·1–1×SSC+50 μg ml^{-1} low molecular weight salmon DNA+0·1% SDS

3×SSC
Kodak X-Omat X-ray film
Ilford fast tungstate intensifying screen

Procedure

Note. Warm all washing solutions to 65°C before use
1   Rinse the hybridization box with solution A.
2   Transfer strips to a soap box and continue rinsing with Solution A until most of the label has been removed.
3   Incubate in three changes of Solution A at 65°C for 40 min.
4   Transfer filters to the final SSC wash mix. The concentration of SSC chosen (between 0·1 and 1×SSC) will determine the homology of hybrids remaining on the filter (see above).
5   Incubate for 1 h at 65°C with one change of solution.
6   Rinse strips with 3×SSC, blot dry, allow to dry in a 37°C room.
7   Reconstruct the filter from the cut strips, tape it to paper sellotaped to a glass plate and cover with Clingfilm or aluminium foil.
8   Cover strips with a sheet of X-ray film.
9   Cover film with intensifying screen followed by another glass plate.
10   Seal the sandwich in at least five layers of aluminium foil, or use a cassette, and autoradiograph overnight at −80°C.

(g) Discusssion of experimental results

The Southern blot filter containing various rabbit and human DNA digests (Table B.1) was hybridized with rabbit β-globin cDNA. The autoradiograph of the hybridized filter is shown in Fig. B.6, together with interpretative maps of rabbit and human β-globin genes.

Rabbit DNA results

When the filters are washed under highly stringent conditions (0·1×SSC at 65°C), only two labelled β-globin DNA fragments are seen in rabbit DNA×*Eco*RI. Since the rabbit adult β-globin gene is cut towards the 3′ end by *Eco*RI, this suggests that each β-globin DNA fragment contains part of the gene and enables a map of *Eco*RI sites neighbouring the gene to be constructed.

   *Pst*I gives one large fragment containing the adult β-globin gene. One *Pst*I site is located in the larger *Eco*RI fragment, which is reduced in size by *Pst*I. This enables the *Pst* fragment to be located relative to the β-globin gene.

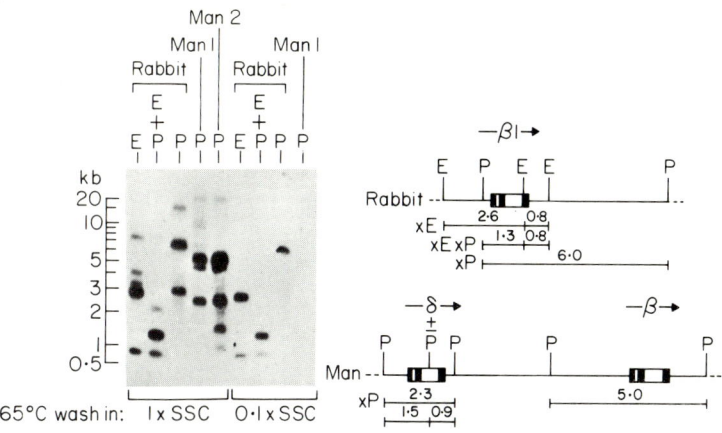

Fig. B.6 DNA fragments detected by rabbit adult β-globin cDNA in restriction endonuclease digests of rabbit and human DNA. Total genomic DNA was digested with *Eco*RI (E) and/or *Pst*I (P) and the blot hybridized with ^{32}P-labelled rabbit β-globin cDNA. After hybridization, filters were given a wash at low stringency (1×SSC at 65°C) or high stringency (0·1×SSC at 65°C) before autoradiography for 2 days at −80°C with a fast tungstate intensifier screen. The physical maps of β-globin genes and neighbouring sequences show the origin of the major hybridizing fragments and show how a map around the rabbit β1-globin gene can be constructed. Human No. 2 is heterozygous for the presence of an extra *Pst*I site (±) within the second intron of the δ-globin gene.

At lower washing stringencies (1×SSC at 65°C), additional fragments containing β-related globin genes are detected. These fragments derive from the rabbit β3- and β4-globin genes and the pseudogene ψβ2. These genes plus the adult β1-globin gene are arranged within a single gene cluster in the rabbit genome in the order 5′-β4-β3-ψβ2-β1-3′ (Lacy *et al.* 1979).

Human DNA results

β-globin gene sequences are conserved in evolution, and consequently the rabbit adult β-globin cDNA probe can detect β-related globin genes in human DNA. At high stringencies, two *Pst*I fragments containing the closely-related human β- and δ-globin genes are detected in individual 1. At low stringencies, additional fragments containing the human ε-, ^{G}γ- and ^{A}γ-globin genes and the pseudogene ψβ1 are also seen (Fritsch *et al.* 1980).

Individual 2 shows two additional *Pst*I fragments not seen in individual 1 and on close inspection can be seen to have a relatively faint δ-globin gene fragment. This person is heterozygous for an additional *Pst*I site within the second intron of the δ-globin gene and has inherited this variant from her mother, who is also heterozygous for the absence/presence of a *Pst*I site within the δ-globin gene (Jeffreys 1979). These restriction fragment polymorphisms are common and provide a major new source of genetic markers in man.

Concluding remarks

Southern blot analysis is a powerful analytical technique that is suitable for estimating the size of gene families, probing gene structure and determining linkage between genes. It can also provide information on the lesions in mutant genes and on genetic variability. It can be used to analyse DNA from any organism, and is also important as a method for locating genes within cloned segments of DNA.

References

Denhardt, D.T. (1966) A membrane-filter technique for the detection of complementary DNA. *Biochem. Biophys. Res. Commun.*, **23**, 641.
Fritsch, E.F., Lawn, R.M. & Maniatis, T. (1980) Molecular cloning and characterization of the human β-like globin gene cluster. *Cell*, **19**, 959.
Jeffreys, A.J. (1979) DNA sequence variants in the $^G\gamma$-, $^A\gamma$-, δ- and β-globin genes of man. *Cell*, **18**, 1.
Jeffreys, A.J. & Flavell, R.A. (1977) The rabbit β-globin gene contains a large insert in the coding sequence. *Cell*, **12**, 1097.
Jeffreys, A.J., Wilson, V., Wood, D., Simons, J.P., Kay, R.M. & Williams, J.G. (1980) Linkage of adult α- and β-globin genes in *X.laevis* and gene duplication by tetraploidization. *Cell*, **21**, 555.
Lacy, E., Hardison, R.C., Quon, D. & Maniatis, T. (1979) The linkage arrangement of four rabbit β-like globin genes. *Cell*, **18**, 1273.
Maniatis, T., Kee, S.G., Efstratiadis, A. & Kafatos, F.C. (1976) Amplification and characterization of a β-globin gene synthesized *in vitro*. *Cell*, **8**, 163.
Rigby, P.W.J., Dieckmann, M., Rhodes, C. & Berg, P. (1977) Labeling deoxyribonucleic acid to high specific activity *in vitro* by nick translation with DNA polymerase I. *J. Mol. Biol.*, **113**, 237.
Schaffner, W., Gross, K., Telford, J. & Birnstiel, M. (1976) Molecular analysis of the histone gene cluster of Psammechinus miliaris: II. The arrangement of the five histone-coding and spacer sequences. *Cell*, **8**, 471.
Smith, H.O. (1980) Recovery of DNA from gels. *Meth. Enzym.*, **65**, 371.
Southern, E.M. (1980) Gel electrophoresis of restriction fragments. *Meth. Enzym.*, **68**, 152.
Wahl, G.M., Stern, M. & Stark, G.R. (1979) Efficient transfer of large DNA fragments from agarose gels to diazobenzyloxymethyl-paper and rapid hybridization by using dextran sulfate. *Proc. natl. Acad. Sci. U.S.A.*, **76**, 3683.

Experiment B.3 Spot hybridizations of animal DNAs

Spot hybridizations on nitrocellulose filters provide a very rapid and sensitive method for estimating nucleotide sequence homologies between various DNAs. In this experiment, DNA samples isolated from various species are denatured and sheared by heating in alkali. The DNAs are then spotted onto a nitrocellulose filter and hybridized with ^{32}P-labelled human DNA. Hybridization to each spot of DNA is estimated by autoradiography.

Materials needed

1 mg ml^{-1} solutions of total genomic DNA from man, brown lemur (a prosimian), rabbit, lion, sheep, frog (*Xenopus laevis*) and *E.coli*

15 µg ml^{-1} total human DNA (for nick translation)

8×3 cm nitrocellulose filter (Sartorius, 0·45 µm pore size)

Procedure

1 Pipette 5 µl of each DNA into an Eppendorf tube.
2 Vacuum dry.
3 Redissolve the DNA in 5 µl 1 M NaCl, 0·1 M NaOH, 10 mM EDTA.
4 Heat for 5 min in a boiling water bath.
5 Spot onto the nitrocellulose filter with spot positions pre-marked with pencil.
6 Leave at room temperature for 30 min.
7 Rinse in 3×SSC for 2 min, blot dry, bake at 80°C for 4 h.
8 Label total human DNA with ^{32}P by nick translation (see Experiment 1, part B).
9 Hybridize the spot filter with ^{32}P-human DNA using the procedures in Experiment 1.
10 Wash out unhybridized label as in Experiment 1, with a final wash in 0·5×SSC, 50 µg ml^{-1} salmon DNA, 0·1% SDS at 65°C.
11 Autoradiograph.

Interpretation of results

The spot hybridization autoradiograph is shown in Fig. B.7.
Under the hybridization conditions used, most labelling is

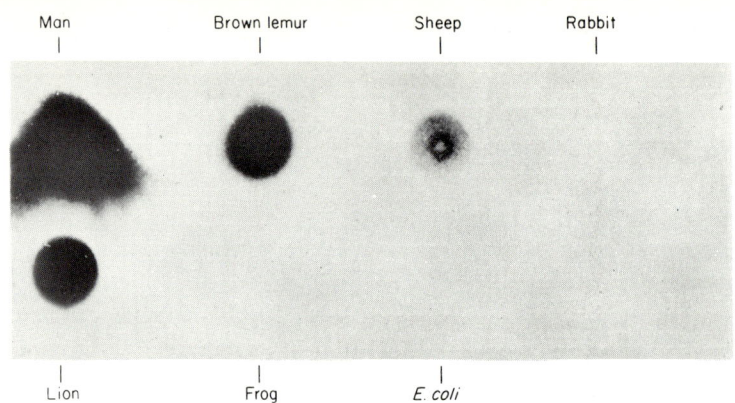

Fig. B.7 Spot hybridization of ^{32}P-labelled human DNA to various species DNAs. The spot filter was autoradiographed overnight at −80°C, using a fast tungstate intensifier screen. The apparent smearing of the human spot is due to the intense labelling of this region of the filter; a shorter exposure shows this spot as a clear labelled circle.

via hybridization of repetitive DNA sequences. As expected, human DNA hybridizes strongly with itself, less well with lemur and weakly with other mammals. Curiously, the lion shows a noticeable homology with human DNA. There is no detectable hybridization with frog or *E.coli* DNA.

The filter washing conditions used (0·5×SSC at 65°C) permit hybrids up to 15% mismatched to be detected. By using more stringent conditions, the stability of hybrids formed with ^{32}P-labelled human DNA may be estimated. In addition, labelled spots may be excised and scintillation counted to quantify levels of cross-hybridization.

This technique is applicable to any set of DNAs and is a useful tool for taxonomic comparisons and for screening cloned DNAs for a sequence of interest.

Experiment B.4 Northern blotting

Background

Northern blotting is a hybridization technique which is used to detect specific RNA sequences in a complex RNA sample. The technique can be applied quantitatively to measure the size (or abundance) of a particular RNA species. The RNA sample is separated on an agarose gel under denaturing conditions and then transferred or blotted to a nitrocellulose filter in the presence of high salt. Baking the filter at 80°C fixes the nucleic acid firmly in position. A radiolabelled probe is incubated with the filter and, after a series of washing steps, the filter is autoradiographed to visualize stable hybrids. Quantitative measurements can be carried out by comparison with appropriate standards. In the experiment described here, a genomic clone coding for the french bean seed storage protein, phaseolin (designated p 7.2, kindly donated by T.C. Hall), is used to detect phaseolin mRNA in various tissues of the plant (Sun *et al.* 1981). It is known that phaseolin is only present in the developing seeds, but it is not known if the gene is transcribed in other tissues.

Preparation of eukaryotic RNA (see Section A.I)

The RNA for this experiment was prepared by the method of Hall *et al.* (1978) but there are many methods available which give RNA of good purity. These methods are reviewed by Taylor (1979). The most important factors during tissue homogenization are the inhibition of ribonuclease and the destabilization of RNA: protein interactions. Ribonuclease is a particular problem when dealing with plant tissues. Destabilization of protein: RNA interactions is usually achieved with EDTA and detergents (e.g. SDS), the latter of which also inhibits ribonuclease. The homogenate is subsequently deproteinized either with redistilled phenol, a mixture of phenol and chloroform or by incubation with a protease. Proteinase K is particularly useful as it is highly active in the presence of SDS. Eukaryotic messenger RNA can be further purified by affinity chromatography on oligo-dT cellulose which binds the polyA tails of mRNA. Further information concerning mRNA purification can be found in section A.1.

Electrophoresis of RNA

RNA is electrophoresed on a horizontal agarose gel under denaturing conditions. Denaturation disrupts secondary and tertiary structure and ensures that mobility is a direct measure of size. Early Northern Blotting experiments were carried out using methyl mercuric chloride as denaturant (Bailey & Davidson 1976), but this compound is extremely toxic and unpleasant to handle. Currently either formaldehyde (Lehrach *et al.* 1977) or glyoxal (McMaster & Carmichael 1977) are used, and both methods have advantages and disadvantages. Formaldehyde is inexpensive and sample preparation is very straightforward. Glyoxal binds irreversibly to nucleic acids at pH 7·0 (which is used for electrophoresis) so a simple phosphate buffer containing no denaturing agent can be used. This confers the advantage that denatured and non-denatured samples can be compared on the same gel. Sample preparation is a little more tedious with glyoxal, as the compound must be deionized before use. Deionized glyoxal can be stored in tightly closed tubes at −20°C but must be discarded after exposure to air. The resolution and sensitivity is good for both techniques.

Blotting

RNA is transferred to nitrocellulose under conditions of high salt essentially as described by Thomas (1980). Early experiments made use of diazobenzyloxymethyl-paper (DBM-paper) which binds RNA covalently (Alwine *et al.* 1980), as it was thought that RNA did not bind tightly to nitrocellulose. In the presence of high salt, binding is perfectly satisfactory. Nitrocellulose is simpler to use than DBM-paper and appears to be at least 10 times more sensitive.

After hybridization and autoradiography, the probe can be removed from the filter by washing, and the filter rehybridized to a second probe. A filter can be recycled in this way several times. Removal of the labelled DNA can be achieved by washing at 65°C for 2 h in 5 mM Tris-HCl, pH 8, 0·2 mM EDTA; 0·05% sodium pyrophosphate and 0·02% each BSA, Ficoll and polyvinylpyrrolidone. After recycling care must be taken not to damage the fragile filter.

References

Alwine, J.C., Kemp, D.J., Parker, B.A., Reiser, J., Renart, J., Stark, G.R. & Wahl, G.M. (1980) Detection of specific RNAs or specific fragments of DNA by fractionation in gels and transfer to diazobenzyloxymethyl paper. *Meth. Enzym.*, **68**, 220.

Bailey, J.M. & Davidson, N. (1976) Methylmercury as a reversible denaturing agent for agarose gel electrophoresis. *Analyt. Biochem.*, **70**, 75.

Hall, T.C., Ma, Y., Buchbinder, B.U., Pyne, J.W., Sun, S.M. & Bliss, F.A. (1978) Messenger RNA for Gl protein of french bean seeds: Cell-free translation and product characterisation. *Proc. natl. Acad. Sci. U.S.A.*, **75**, 3196.

Lehrach, H., Diamond, D., Wozney, J.M. & Boedtker, H. (1977) RNA molecular weight determinations by gel electrophoresis under denaturing conditions, a critical reexamination. *Biochemistry*, **16**, 4743.

McMaster, G.K. & Carmichael, G.C. (1977) Analysis of single- and double-stranded nucleic acids on polyacrylamide gels by using glyoxal and acridine orange. *Proc. natl. Acad. Sci. U.S.A.*, **74**, 4835.

Sun, S.M., Slightom, J. & Hall, T.C. (1981) Intervening sequences in a plant gene-comparison of the partial sequence of cDNA and genomic DNA of french bean phaseolin. *Nature*, **289**, 37.

Taylor, J.M. (1979) The isolation of eukaryotic messenger RNA. *Ann. Rev. Biochem.*, **48**, 681.

Thomas, P.S. (1980) Hybridisation of denatured RNA and small DNA fragments transferred to nitrocellulose. *Proc. natl. Acad. Sci. U.S.A.*, **77**, 5201.

(a) Electrophoresis of RNA

Materials needed

1 mg ml^{-1} developing seed RNA

1 mg ml^{-1} seedling RNA

1 mg ml^{-1} mature leaf RNA

Marker DNA (as for Southern blot)

Agarose beads as for DNA gel (as Appendix B.II but agarose dissolved in 2·2 M formaldehyde; 20 mM Na phosphate, pH 7·6; 10% glycerol plus bromophenol blue)

The simple gel electrophoresis apparatus used in these experiments is illustrated in Fig. B.8.

Procedure

Prepare slab gel

1 Dissolve 1 g agarose in 50 ml sterile distilled water by boiling in a 250 ml conical flask.

2 In a universal bottle, dissolve 0·1 g agarose in 10 ml sterile water by boiling.

3 Seal the mesh-covered openings of the plate with masking tape and set down on a level surface (Fig. B.8).

4 Suspend a slot former near one end of the plate as described for the DNA gel.

5 From the universal bottle, pipette 5 ml of agarose into the well at each end of the plate.

6 Allow to set (about 10 min) then rapidly add 50 ml 2×gel buffer (4·4 M formaldehyde; 50 mM Na phosphate, pH 7·6; 1 μg ml^{-1} ethidium bromide) to the conical flask and swirl to mix.

7 Pour on to plate and allow to set (about 45 min).

8 Place gel into apparatus, after removing comb, and add enough electrophoresis buffer (2·2 M formaldehyde; 10 mM Na phosphate, pH 7·6) to come 0·5 cm up the 'legs' of the gel.
9 Fill the sample slots up with electrophoresis buffer.

Note. All manipulations involving formaldehyde must be carried out in a fume hood.

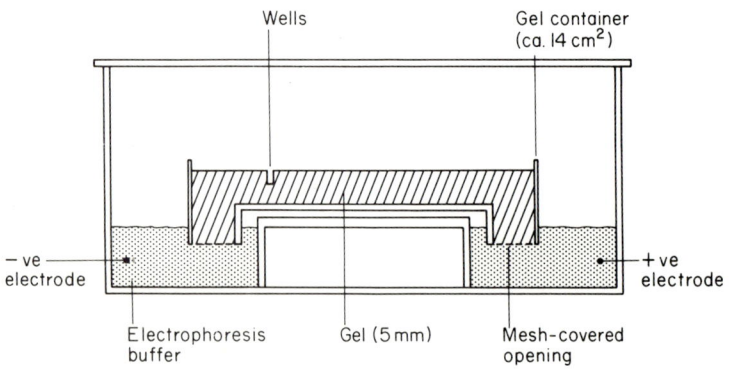

Fig. B.8 A simple apparatus for gel electrophoresis. This apparatus can be easily made in perspex using plastic mesh to support the gel 'legs' or wicks. It is useful for running non-submerged gels which require a much lower current than submerged gels. An inexpensive power pack will provide this low current. RNA gels run satisfactorily under either condition.

RNA samples

1 Five microlitres of each RNA sample is vacuum-dried and redissolved in 10 μl denaturation mix (50% formamide; 2·2 M formaldehyde; 20 mM Na phosphate, pH 7·6; 0·05% bromophenol blue).
2 Denature at 60°C for 10 min and add 10 μl agarose beads and load into slots on agarose gel (see Table B.3 for loading pattern) along ith 10 μl alkaline denatured marker DNA.
3 Switch on power pack at 75 V.
4 Using a cut-off Pasteur pipette, carefully pipette some melted vaseline over the sample wells. This prevents the wells drying out during electrophoresis, which can cause trailing of the RNA.
5 Run for about 1·5 h until the bromophenol blue has migrated about 8 cm.

Table B.3 Loading pattern for gel

R1 alkali denatured marker DNA—see Southern blot experiment (Experiment B.2)
R2 —
R3 seed RNA
R4 seedling RNA
R5 mature leaf RNA
R6 —
R7 alkali denatured marker DNA

(b) Preparation of ^{32}P-labelled probe by nick translation of DNA

See Experiment B.2 Nick translation.

(c) Blotting the RNA agarose gel

Procedure

1   Using a razor blade cut the gel out of the mould by running the blade along the edge of the gel and then cutting the gel to leave the 'legs' behind in the plate. Transfer the gel to water at 60°C for 10 min. Very carefully pour off the hot water and add cold water. Leave for 10 min. These steps drive some of the formaldehyde out of the gel.
2   View under ultraviolet light.
3   Wash gel for 20 min in 20×SSC.
4   Set up transfer apparatus.
This is done exactly as described for the Southern blot experiment (Experiment B.2) except that the nitrocellulose and 3 MM filters are soaked in 20×SSC. The nitrocellulose filter should be wetted with sterile water before placing in 20×SSC; this step aids even wetting of the nitrocellulose. Leave to transfer overnight.

(d) Baking the RNA nitrocellulose filter

Procedure

1   Remove the filter from the gel, marking the uppermost side with pencil.
2   Blot dry on Whatman 3 MM paper.
3   Bake in vacuum oven at 80°C for 2 h.

(e) Hybridization of the RNA filter with ^{32}P-labelled DNA

Materials needed

Pre-hybridization buffer (50% v/v formamide; 50 mM Na phosphate, pH 6.5; 5×SSC; 250 μg ml^{-1} denatured salmon sperm DNA; 0·02% BSA; 0·02% Ficoll (Sigma, type 400); 0·02% PVP (Sigma PVP-360); 25 μg ml^{-1} polyadenylic acid)
Hybridization buffer, (four parts of the above, but with polyadenylic acid omitted, mixed with 1 part 50% dextran sulphate)
2×SSC+0·1% SDS
1×SSC+0·1% SDS
0·1×SSC+0·1% SDS

Procedure

1 Cut the filter for the DNA filter according to the patterns in Table B.4. The marker DNA strips are hybridized to ^{32}P-DNA as described in Experiment B.2.

2 Prehybridized RNA filters at 42°C for 8–20 h in pre-hybridization buffer. In this experiment the time available will limit this procedure to about 7 h. This step can be carried out in soap boxes whilst the subsequent hybridization is carried out in the hybridization chamber exactly as for DNA hybridization.

3 Denature nick translated probe by immersing in a 100°C bath for 5 min and added to the hybridization mix in the chamber. Mix thoroughly.

4 Transfer the RNA containing filters to the hybridization buffer, touching on to dry 3 MM paper to remove excess pre-hybridization mix. Incubate at 42°C overnight.

Table B.4 Cutting pattern for DNA filter

Sample number	Probe	Box number	Posthybridization wash
R1		1	1×SSC
R2 R3 R4 R5 R6	p 7·2	3	See below
R7		1	1×SSC

(f) Washing the RNA filters

Procedure

1 The RNA blots are transferred to a box and washed in four changes of 2×SSC, 0·1% SDS for 10 min at room temperature, then with 1×SSC, 0·1% SDS followed by two changes of 0·1 SSC for 15 min at 50°C. Blot the strips dry and allow to dry in a 37°C room.

2 Reconstruct the filter from the strips and tape it to paper sellotaped to a glass plate. Cover with Clingfilm.

3 Expose to X-ray film exactly as described for the Southern blot (experiment B.2). Leave at −70°C overnight.

Appendix B.I Additional notes and trouble-shooting for experiments in Section B

Restriction endonuclease digestions

Occasionally, a DNA sample is encountered which cannot be cleaved with a restriction endonuclease. This can occur for example when DNA is prepared from a carbohydrate-rich source without any steps to remove carbohydrate (see Appendix D.II), or when DNA is recovered from an agarose gel. This inhibition can usually be overcome by adding spermidine trichloride to the reaction to a final concentration of 4 mM at 0°C immediately before adding the restriction enzyme. Spermidine also helps with ligations, nick translations, fill-in labelling and kinase labelling. However, take care if using spermidine with high molecular weight DNA in a low salt buffer as DNA precipitation can occur.

Southern blotting

In the experiment described in this Section the restricted DNA was denatured with alkali before electrophoresis in an agarose gel. This procedure has several advantages over native gels.

Advantages

More DNA can be loaded without overloading the gel.
A high % agarose gel can be used; these are easier to handle during the blotting procedure.
Large fragments are better resolved than on a native gel.
Once run, the DNA is single stranded and ready for transfer.

Disadvantages

DNA bands tend to be broader than on native gels, and are sometimes curved due to salt effects in the slot.
DNA must be relatively free from nicks.
Restriction endonucleases must be free from nicking activity (they almost always are).
The sizes of large DNA fragments tend to be overestimated.

If double-stranded DNA is electrophoresed, the DNA in the agarose gel must be denatured *in situ* after electrophoresis by soaking the gel in NaOH and neutralizing in NaCl/Tris-HCl (Southern 1980).

Large DNA fragments tend to be transferred inefficiently. This problem can be avoided by the procedure of Wahl *et al.* (1979) in which the gel is soaked in dilute HCl to depurinate the DNA partially, then soaked in alkali, to nick and denature DNA, neutralized and blotted.

Small DNA fragments (< 300 base pairs) tend not to bind efficiently to nitrocellulose. This difficulty can be avoided by blotting onto DBM paper (Alwine *et al.* 1979) to which DNA becomes covalently attached.

Southern blots on nitrocellulose filters can be re-used by removing the original hybridized probe, by extensive washing of filters in five changes of water at 65°C for 1·5 h. After this treatment, the filters can be rehybridized using the usual procedure. We have taken filters through six cycles of hybridization, although on the final hybridization, the signal is reduced to about ⅓ of the initial signal (Jeffreys *et al.* 1980).

Northern blotting

To save time, the ethidium bromide has been added to the gel in this experiment. However, better staining is achieved by omitting it from the gel and staining the gel afterwards. To achieve this, after the gel has been soaked in hot and cold water, it is transferred to 1 μg ml^{-1} ethidium bromide for 30 min and then destained in water for 30–60 min.

References

Alwine, J.C., Kemp, D.J., Parker, B.A., Reiser, J., Renart, J., Stark, G.R. & Wahl, G.M. (1979) Detection of specific RNAs or specific fragments of DNA by fractionation in gels and transfer to diazobenzyloxymethyl paper. *Meth. Enzm.*, **68**, 220.

Jeffreys, A.J., Wilson, V., Wood, D., Simons, J.P., Kay, R.M. & Williams, J.G. (1980) Linkage of adult α- and β-globin genes in *X. laevis* and gene duplication by tetraploidization. *Cell*, **21**, 555.

Southern, E.M. (1980) Gel electrophoresis of restriction fragments. *Meth. Enzm.*, **68**, 152.

Wahl, G.M., Stern, M. & Stark, G.R. (1979) Efficient transfer of large DNA fragments from agarose gels to diazobenzyloxymethyl-paper and rapid hybridization by using dextran sulfate. *Proc. natl. Acad. Sci. U.S.A.*, **76**, 3683.

Appendix B.II   Electrophoretic separation of DNA fragments in agarose gels

Electrophoresis tank

A simple and inexpensive perspex chamber suitable for running submerged agarose slab gels is shown in Fig. B.9. This chamber can be used with any power-pack capable of delivering up to 500 V/500 mA.

Preparation of slab gels

Agarose is dissolved by boiling in electrophoresis buffer (40 mM Tris-acetate, pH 7·7; 1 mM EDTA; 0·5 µg ml^{-1} ethidium bromide) to 0·4–3% agarose (w/v). The boiling is most conveniently carried out in a microwave oven.

A gel mould is made by wrapping scotchtape around the edge of a glass plate (20×20 cm or smaller) to form a wall. A perspex slot former 1·5 mm thick thick with 5 mm wide teeth is clamped with bulldog clips vertically about 2 mm above, and near one end, of the plate. The scotchtape walls are supported by lumps of plasticine. The boiled agarose solution is quickly cooled to about 50°C (for < 1% gels), poured onto the plate to a depth of 6–10 mm and allowed to set for at least 1 h. The gel surface is dampened with electrophoresis buffer, the slot former and scotchtape removed and the gel together with its glass plate placed in the horizontal gel tank with enough electrophoresis buffer to cover the gel.

Preparation and electrophoresis of DNA samples

DNA samples are prepared either: (i) by mixing with 1/5 vol of sample buffer (0·1% bromophenol blue, 20% w/v glycerol) prior to loading or (ii) by mixing with 1 vol of agarose beads prior to loading.

Agarose beads are prepared by dissolving agarose to 0·2% w/v by boiling in 10 mM Tris-HCl (pH 7·5), 20 mM EDTA, 10% glycerol plus bromophenol blue. The agarose is allowed to set at room temperature and, once set, is forced several times through a syringe plus needle to give a fine slurry of agarose beads which can be kept indefinitely at 4°C. These beads aid loading and help prevent electrophoretic trailing of DNA at the sides of the sample slots (Schaffner *et al.* 1976).

Gels are electrophoresed at 10 V/cm for 1–2 h. The gels are

Fig. B.9 Gel tank. Overall dimensions of the tank are: length 30 cm, width 20 cm, depth 10 cm.

then viewed and photographed on a short wave UV-Birchover Spectro-light Transilluminator (**Care!**) using a Polaroid camera fitted with suitable ultraviolet filters.

Choice of agarose gel concentration

This depends on the size of DNA fragment being resolved. 0·5% gels are suitable for fragments in the 1–20 kb range, and 2% gels for 0·1–1 kb fragments.

Re-use of gels

Agarose slab gels can be re-used indefinitely, provided that the electrophoresis tank is replenished with fresh electrophoresis buffer and that the slots are rinsed with buffer. Take care during a long electrophoresis run, as the ethidium bromide is bleached in the anodal compartment and, once depleted, will result in the gel being electrophoretically destained.

Appendix B.III Recovery of DNA from agarose gels

Run sample in 0·3–3·0% (depending on fragment size) agarose + 0·5 μg ml^{-1} ethidium bromide. Visualize using a longwave overhead ultraviolet lamp and cut out the required band using a scalpel. Keep ultraviolet exposure to a minimum to avoid damage to DNA.

(a) Electroelution method

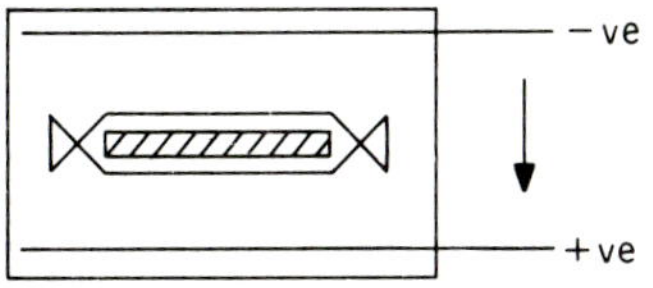

1 Place gel slice in dialysis bag and just cover the slice with electrophoresis buffer. Tie and seal bag expelling air bubbles.
2 Replace bag in electrophoresis tank, submerged, and arrange so that DNA will come out of slice by shortest route (Fig. B.10). Turn on power to 50 V and check occasionally by overhead ultraviolet. DNA should collect *on* the dialysis bag, nearest the anode. This may take several hours.
Care! Live terminals exposed.
3 When all DNA is collected on dialysis bag, turn off current, remove bag from tank, agitate gently to resuspend DNA in buffer, then collect this buffer taking care *not* to collect pieces of agarose as well.
4 Extract with butan-2-ol to remove ethidium bromide and also to concentrate the solution. Ether extract to remove butanol and collect DNA by ethanol precipitation.

Fig. B.10 Gel slice in dialysis bag submerged in electrophoresis tank.

(b) Electrophoresis onto DEAE-cellulose paper
(Dretzen *et al.* 1981)

1 Whatman DE81 paper is prepared by cutting into 5 × 3 cm strips, soaking in 2·5 M NaCl for 3 h and washing 5 times with water. The strips are stored in 1 mM EDTA, pH 8·0, at 4°C.
2 Wrap the agarose gel slice in a strip of DE81 paper and insert the package into a slot in a horizontal agarose gel. Electrophorese at 20 V cm^{-1} for 20 min.
3 Place the package in water in a Petri dish, unwrap and discard the water. Locate the region of paper-bound DNA under longwave ultraviolet, cut out with scalpel, wash with water and blot dry.
4 Place the piece of DE81 paper in an Eppendorf tube and add 0·45 ml 1 M NaCl, 50 mM Tris-HCl (pH 7·5), 1 mM EDTA.

Shake vigorously to disperse the paper, and incubate at 37°C for 15 min.

5 Invert the tube, puncture the bottom of the tube with a hot, fine needle and centrifuge within a second centrifuge tube to collect the DNA solution free from DE81 paper.

6 Add 1·1 ml ethanol, mix, chill for 5 min in an ethanol/dry-ice bath, and collect the DNA by centrifuging in an Eppendorf centrifuge for 5 min. Rinse the DNA pellet with 70% ethanol, vacuum dry and dissolve in water.

References

Dretzen, G., Bellard, M., Sassone-Corri, P. & Chambon, P. (1981) A reliable method for the recovery of DNA fragments from agarose and acrylamide gels. *Anal. Biochem.*, **112**, 295.

Schaffner, W., Gross, K., Telford, J. & Birnstiel, M. (1976) Molecular analysis of the histone gene cluster of *Psammechinus miliaris. Cell*, **8**, 471.

Section C   DNA-dependent gene expression systems

Julie Pratt
Barry Holland
Neil Stoker

Introduction

Once a specific gene or linked array of genes has been cloned into a suitable vector system it is often important to be able to identify the corresponding polypeptide products. Four major prokaryote gene expression systems are currently available:

1 **Ultraviolet (UV)-irradiated host system.** *E. coli* is heavily UV-irradiated to minimize host protein synthesis and then infected with phage carrying cloned genes. ^{35}S-methionine is included in the subsequent incubation to label phage-encoded proteins and these products are analysed by SDS-PAGE. As an extra sophistication, synthesis of the many λ-proteins may be blocked by the use of a host lysogenic for λ or carrying a plasmid coding for the λ transcriptional repressor cI.

2 **Maxi-cell system.** Where genes can be successfully cloned on to a high copy number plasmid either the maxi-cell or mini-cell system may be used. After transformation with the relevant multicopy plasmid, the maxi-cells (usually *rec*A *uvr*A to ensure extreme irradiation sensitivity) are subjected to UV-irradiation. After 16 h of incubation, the chromosome is largely destroyed but a few copies of the plasmid will remain intact. Subsequent incubation with ^{35}S-methionine results in specific labelling of proteins coded by the plasmid*.

3 **Mini-cell system.** In the case of the mini-cell system, transformation of a *min*A *min*B strain of *E. coli* (which produces large numbers of anucleate mini-cells) with a multicopy plasmid results in the formation of mini-cells into which plasmid DNA has segregated but not chromosomal DNA. Labelling of purified mini-cells then provides an excellent system for the specific identification of plasmid-coded proteins*.

4 *In vitro* **transcription/translation system.** The *in vitro* synthesis of proteins programmed by either phage or plasmid DNA in a coupled transcription/translation system (Zubay) constitutes perhaps the most versatile method. Direct manipulation of the DNA *in vitro* before its use as a template

---

*Note. Large plasmids (40 kb or more) which are usually low copy number (i.e. 1–3 cell^{-1}) can also be analysed using systems 2 and 3, but with lower efficiency. In the case of maxi-cells the larger the plasmid the greater the probability of its receiving some DNA damage after UV-irradiation and in the case of mini-cells the chance of a large, low copy number plasmid segregating into a mini-cell is much lower.

can pin-point relatively small regions of DNA coding for a specific polypeptide.

The objective of the experiments in this section is to demonstrate the use of each of these four gene expression systems. The most appropriate system for a particular use will depend on the cloning system that has been adopted. In addition, the different systems have their respective merits and demerits, and these are compared in Table C.1.

Table C.1 Comparison of the four gene expression systems

System	Advantages	Limitations
λ infected UV-irradiated hosts	Simple procedure, relatively easy to set up. The powerful pL-λ promoter can be harnessed in certain cases to enhance expression of weakly expressed genes. Cloning into λ initially can be essential for genes whose presence in more than a single copy may be lethal *in vivo*. Has marked advantages over the mini-cell system when screening through large numbers of extension or deletion families of phages to identify gene orders in cloned sequences	Clearly limited to λ-vectors. Residual background synthesis of host polypeptides below 25 kDa can obscure identification of cloned gene products in this size range. System makes relatively inefficient use of the labelled amino acids supplied, therefore may require long autoradiographic exposure times and/or fluorography. In our hands we also find that repression of phage protein synthesis by introduction of a 'cI' repressor into the irradiated host is often incomplete with several variants of λ
Maxi-cells	Simple procedure, relatively clean background, less susceptible to problems of viable cell contamination than mini-cells	Fairly efficient system for isotope labelling. As with the mini-cell method, this system does not readily distinguish between genes expressing from their own promoter or from a promoter in the vector
Mini-cells	Relatively clean background to gel profiles with virtually no expression of host genes. High fidelity of product formation since the system is essentially *in vivo*. In principle can be used both for plasmids and λ infection of purified mini-cells but the latter use has not been widely adopted	High copies of certain genes (e.g. coding envelope proteins) when present on plasmids in the mini-cell-producing strain can severely disturb mini-cell formation and/or efficient purification of mini-cells. Fairly efficient system for isotope labelling
In vitro (Zubay)	Can be used with any kind of cloned DNA including small linear fragments. This allows unambiguous assignment of cloned DNA products (rather than those coded by the vector) to relatively small regions of DNA. Extremely efficient labelling system allowing rapid analysis. Can also be used to study gene expression and has considerable advantages when uncharacterized gene products must be identified	Relatively complex system to set up initially. Interpretation can sometimes be obscured by the presence of some abortive translation products and/or proteolytic breakdown products especially in the case of large proteins > 60 kDa.

Experiment C.1 λ infection of ultraviolet-irradiated cells

Background

In this system, transcription of the bacterial chromosome is prevented by administration of a large dose of UV-irradiation. In a subsequent infection with λ, only the phage DNA should be a functional template for transcription, and addition of ^{35}S-methionine will allow phage-encoded proteins to be specifically labelled.

If the bacterial strain which is irradiated is lysogenic for λ, then it will contain the repressor protein (the cI gene product). This repressor will bind to any superinfecting phage, preventing all transcription from the major promoters, pL and pR. Only the immunity region, which encodes two proteins—the repressor itself, and the rex gene product—is therefore transcribed together with any bacterial genes cloned into the

Table C.2 Plasmids used in Section C

		Size (kb)	Copy number (per genome)	Amplifiable	Comments
(A)	pACYC184	4·0	18	Yes	
(B)	pLG517	10·4	18	Yes	6·4 kb *Eco*RI fragment with *dac*C gene in Cm gene of pACYC184
(C)	pBS42	12·2	6	No	1·64 kb *Bam*HI-*Eco*RI fragment with *dac*A
(D)	pBR325	6·0	50	Yes	

Bacteriophage
| (E) | λ+ | | | | |
| (F) | λ *dac*C | | | | 6·4 kb *Eco*RI fragment cloned into a λ cloning vector |

Gene products encoded by plasmids and phage

Cm^R	Chloramphenicol acetyl transferase (CAT); a 25 kd cytoplasmic protein
Km^R	Kanamycin phosphotransferase; a 27 kd cytoplasmic protein
Amp^R	TEM-1 β-lactamase; a 29 kd protein found in the periplasmic space; made as a 31·5 kd precursor
Tc^R	Tetracycline resistance protein; a 34 kd inner (cytoplasmic) membrane protein
PBP5 (*dac*A)	Penicillin-binding protein 5; a 42 kd inner membrane protein with D-alanine carboxypeptidase activity involved in the synthesis of the peptidoglycan layer; made as a 45 kd precursor
PBP6 (*dac*C)	Penicillin-binding protein 6, a 40 kd inner membrane protein with carboxy-peptidase activity, related to PBP5; made as a 44 kd precursor
cI	The λ repressor protein (29 kd)
rex	A 27 kd protein with unknown function, encoded by the λ immunity region

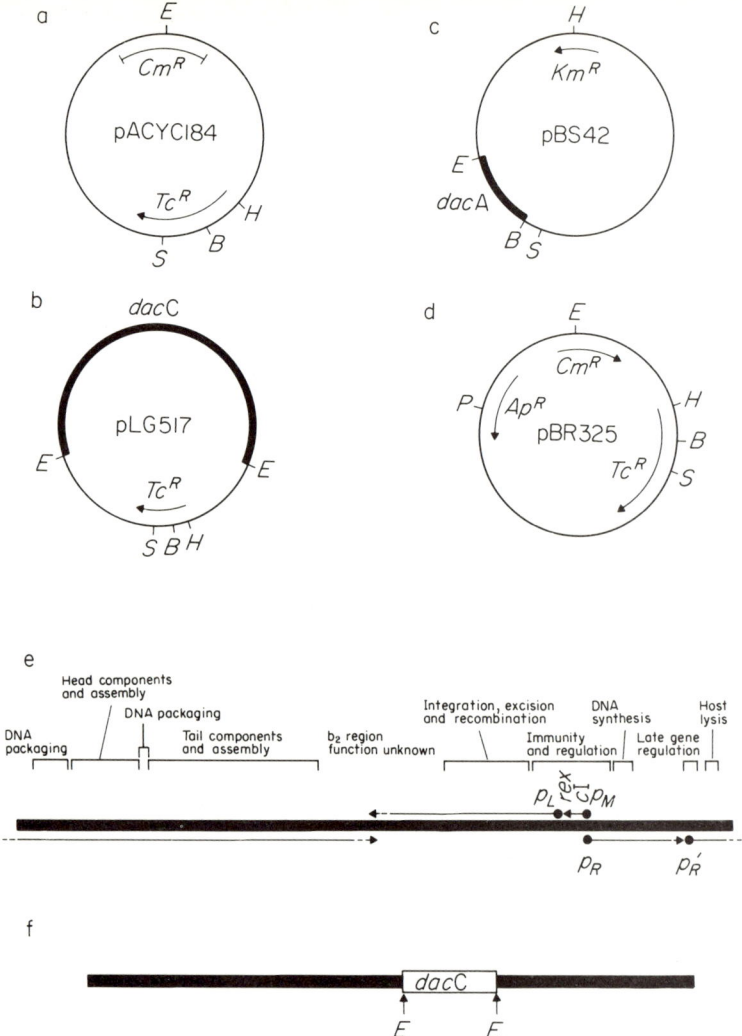

Fig. C.1 Plasmids and phages listed in Table C.2.

superinfecting phages which have their own promoters. The bacterial strains to be used will be *E. coli* 159 (*uvr*A) and the same strain lysogenized with λ *ind*. UV-irradiation normally induces excision of a prophage but λ *ind* carries a mutation which prevents this.

Superinfecting λ can only enter the cell if the lamB gene product is present in the outer membrane. This protein, normally involved in maltose transport, acts as a receptor for the phage. It is normally only present in a few molecules per cell, but can be induced in much greater amounts by growth of the bacteria in maltose.

The phages used in this experiment will be wild-type λ, and λ *dac*C carrying a 6·4 kb *Eco*RI fragment containing the *dac*C gene, encoding PBP6 (see Fig. C.1 and Table C.2).

Phage suspensions are dialysed extensively to remove any methionine so as to increase the efficiency of incorporation of ^{35}S-methionine.

The objectives of the experiment are:
1. To demonstrate the use of the UV-irradiated host system.
2. To show the effect of λ repressor protein on transcription.
3. To identify the cI and rex gene products of λ.
4. To identify the gene products encoded by the *Eco*RI fragment cloned into λ *dac*C.

λ infection of UV-irradiated hosts

Materials needed

Sterile distilled water
M9 salts*
CaMg salts*
20% maltose
Lambda buffer (LB)*
Methionine (8 mg ml^{-1})
^{35}S-methionine
Buffer B*
SDS-sample buffer†
0·1 M MgCl$_2$
Bacterial strains
   159; *uvr*A
   159: *uvr*A (λ*ind*)
Bacteriophage
   λ$^+$
   λ *dac*C

Table C.3 Incubations required for Experiment C.1

159 + LB (control)
159 + λ$^+$
159 + λ *dac*C
159 (λ *ind*) + LB (control)
159 (λ *ind*) + λ$^+$
159 (λ *ind*) + λ *dac*C

Procedure

1   Make up M9-minimal medium as follows, using sterile technique:

Sterile distilled water	100 ml
M9 salts	10 ml
Ca/Mg salts	1 ml
20% maltose	1 ml

Mix, supplement with 10 mM MgCl$_2$ and place in fridge.

2   Use 10 ml to inoculate with one colony of 159 and 159 (λ *ind*). Place in a shaking water bath at 37°C and incubate overnight.

---

*See Appendix C.I
†See Appendix C.II.

3 Take 1 ml of each culture and inoculate 25 ml of minimal medium. Incubate at 37°C in shaking water bath until an A_{450} of 0.64 (about 2×10^8 cells ml^{-1}) is reached. Cultures with A_{450} 0.4–0.8 can be used if volumes are adjusted to give 2×10^8 cells in total.

4 Pipette 3.5 ml into a small disposable Petri dish and UV-irradiate with swirling (1,200 Jm^{-2}).

5 Pipette 500 µl of each irradiated culture into 3 Eppendorf tubes and pellet the cells by centrifugation (2 min).

6 Resuspend the cells in 100 µl pre-warmed minimal medium containing 10 mM $MgCl_2$ and add 100 µl phage, or lambda buffer, as indicated in Table C.3. The phage concentration should be 10^{10} ml^{-1} to give a multiplicity of infection of about 10.

7 Incubate at 37°C for 10 min to allow phage to adsorb.

8 Add 200 µl prearmed M9 medium to each tube, and incubate at 37°C for 20 min to allow host mRNA to decay.

9 Add 5 µl ^{35}S-methionine (approximately 60 µCi) to each tube. Incubate for 10 min at 37°C.

10 Add 200 µl unlabelled methionine (8 mg ml^{-1}) to each tube, and incubate for 5 min at 37°C.

11 Chill the samples on ice for 2 min.

12 Pellet the cells (2 min, Eppendorf centrifuge) and discard the supernatant.

13 Resuspend the pellet in 0.5 ml buffer B and harvest again.

14 Resuspend in 15 µl buffer B, and add 15 µl SDS-sample buffer.

15 Heat the samples for 3 min in a boiling water bath, and load the entire sample onto a SDS-polyacrylamide gel, prepared as instructed and as described in Appendix C.II.

Trouble shooting

1 The optimum multiplicity of infection of λ is between 5 and 10. A higher phage:bacterium ratio may result in the cI repressor being titrated out, allowing phage genes to be transcribed. In the case of λ, a successful run is indicated by the expression of only two major phage proteins cI and rex. A lower multiplicity of infection results in lower expression of cloned genes.

2 The phage must be extensively dialysed against λ buffer to remove endogenous methionine, otherwise incorporation of added label is severely reduced.

3 It may be necessary to optimize the UV dose to obtain a low background in the controls without added phage.

4 Many λ cloning vectors carry the phage 434 or 21 immunity regions. For repression of the phage genes in these cases 159

(imm434) or 159 (imm21) should be used respectively. We find, however, that these prophages do not repress the superinfecting phages as efficiently as the λ ind prophage. When working with phages carrying 434 immunity, we therefore use 159 (pGY101). pGY101 is a plasmid into which the 434 repressor gene has been cloned, and produces approximately 70 times as much repressor protein as a λ imm434 prophage (Levine et al. 1979).

5 Proteins synthesized from λ transducing phage can also be detected by infection of a UV-irradiated non-lysogen of $E. coli$. In this case the absence of λ repressor in the cells results in the transcription of phage genes from pL and pR in addition to transcription from chromosomal promoters on the transducing phage. This may have the advantage that it may allow the expression of chromosomal genes on the phage from the powerful λ promoters. This can be useful for chromosomal genes that have weak promoters or have been separated from their promoters. It can also be used to determine the direction in which a gene is transcribed (Ward & Murray 1979, Lutkenhaus & Wu 1980).

References

Levine, A., Bailone, A. & Devoret, R. (1979) Cellular levels of the prophage λ and 434 repressors. $J. Mol. Biol.$, **131**, 655.

Lutkenhaus, J.F. & Wu, H.C. (1980) Determination of transcriptional units and gene products from the $ftsA$ region of $Escherichia coli$. $J. Bacteriol.$, **143**, 1281.

Spratt, B.G., Boyd, A. & Stoker, N. (1980) Defective and plaque-forming lambda transducing bacteriophage carrying penicillin-binding protein-cell shape genes: Genetic and physical mapping and identification of gene products from the lip-$dacA$-$rodA$-$pbpA$-$leuS$ region of the $Escherichia coli$ chromosome. $J. Bacteriol.$, **143**, 569.

Ward, D.F. & Murray, N.E. (1979). Convergent transcription in bacteriophage λ: Interference with gene expression. $J. Mol. Biol.$, **133**, 249.

Experiment C.2 Maxi-cells

Introduction

E. coli contains several systems for repairing DNA damaged by UV-irradiation and some other mutagens. Mutations in the *rec*A and *uvr*A genes inactivate the major repair systems, making cells extremely sensitive to UV-irradiation. The damaged DNA is then degraded by nucleases and subsequent incubation of these so-called maxi-cells results in very little protein synthesis due to the lack of transcribable DNA. However, if such strains carry a multicopy plasmid, many plasmid molecules will survive the UV treatment because they are a much smaller target than the chromosome and will consequently serve as templates for transcription in subsequent incubations. The addition of labelled amino acids under these conditions therefore leads to differential labelling of polypeptides coded by the plasmid.

Background

E. coli CSR603 is a strain which is often used to produce maxi-cells. It is deficient in all the major repair pathways, since it carries *rec*A, *uvr*A and *phr* mutations. It is therefore extremely sensitive to UV, a single UV-induced lesion in the chromosome being lethal. This strain is easily transformed with multicopy plasmids, and large plasmids can be conjugated into it. However, it is sometimes difficult to maintain; survival on plates is poor, and in our hands the background incorporation (due to residual host templates) is rather variable. We have therefore used an alternative, *E. coli* strain CSH26 *rec*A.

The *rec*A gene in this strain has been deleted. While being less sensitive to UV than CSR603, intelligent handling can lead to the production of maxi-cells with background incorporation as low as for CSR603. CSH26 *rec*A has the advantage of being a strain generally easier to grow and to maintain. Many cloning experiments normally involve transformation into a *rec*A strain at an early stage, in order to minimize any rearrangements of the cloned DNA sequences. Thus the same strain should be capable of serving as a maxi-cell, facilitating the identification of the cloned gene products, without the need to transfer the plasmid to a specific 'maxi-cell' host.

The plasmids pACYC184, pLG517 and pBR325 (see Table C.2) have been transformed into both CSR603 and CSH26 recA and in this experiment the objectives are:

1 To compare the suitability of CSR603 and CSH26 recA as maxi-cell strains.
2 To identify the gene products encoded by the plasmids pACYC184, pLG517 and pBR325.
3 To compare the results with those obtained using mini-cells and the Zubay system (Experiments C.3 and C.4).

Materials provided

K medium*
Hershey salts*
Hershey medium*
Streptomycin sulphate (100 mg ml^{-1})
Tetracycline hydrochloride (10 mg ml^{-1})
Sodium ampicillin (1·5 mg ml^{-1})
Cycloserine (4 mg ml^{-1})
Difco methionine assay medium (10·5%)
Methionine (8 mg ml^{-1})
^{35}S-methionine
Buffer B*
SDS-sample buffer†

Bacterial strains
    CSR603:   *phr*1   *rec*A1   *uvr*A6   *thr*1   *leu*B6   *arg*E3
                    *thi*1   *ara*14   *lac*Y1   *gal*K2   *xyl*5   *mtl*1
                      *sup*E44   *tsx*33   *gyr*A96   *rps*L31
    CSH26 ΔF6: F$^-$ *ara thi* Δ(*lac pro*) Δ*recA srl*) *rps*L
    CSR603 [pACYC184]   CSH26 *rec*A [pACYC184]
    CSR603 [pLG517]   CSH26 *rec*A [pLG517]
    CSR603 [pBR325]   CSH26 *rec*A [pBR325]

Procedure

1 Put up eight overnight cultures as in Table C.4.
2 Pipette 10 ml K medium into labelled test tubes, add streptomycin and tetracycline (10 µl of each stock) as appropriate and 0·5 ml overnight culture. Incubate at 37°C in a shaking waterbath to A_{450} 0·5.
3 Transfer 3.5 ml of each culture into small disposable Petri dishes, and UV-irradiate (with swirling) with 1·36 J m^{-2}.

*See Appendix C.1.
†See Appendix C.II.

Table C.4

Strain	Plasmid	Antibiotic to be added
CSR603	—	str
CSR603	pACYC184	str, tet
CSR603	pLG517	str, tet
CSR603	pBR325	str, tet
CSH26 recA	—	str
CSH26 recA	pACYC184	str, tet
CSH26 recA	pLG517	str, tet
CSH26 recA	pBR325	str, tet

4 Pipette 3 ml of the irradiated culture into a sterile test-tube, and cover the tubes with silver foil. Incubate at 37°C for 1 h, with shaking.

5 Add 60 µl ampicillin (1·5 mg ml^{-1}; final concentration = 30 µg ml^{-1}) to each culture except the one carrying pBR325 (which encodes β-lactamase). To this one add 75 µl cycloserine (4 mg ml^{-1}; final concentration = 100 µg ml^{-1}). Incubate overnight at 37°C to allow degradation of chromosomal DNA.

6 Transfer 0·5 ml of each maxi-cell culture to a sterile Eppendorf tube.

7 Harvest the cells (3 min, Eppendorf centrifuge), discard the supernatant, and wash the pellet twice with Hershey salts.

8 Resuspend the cells in 200 µl Hershey medium and add 4 µl ampicillin (or 5 µl cycloserine to the pBR325-containing strain). Incubate at 37°C for 1 h.

9 Add 2 µl ^{35}S-methionine (25 µCi) and 3 µl Difco methionine assay medium. Incubate at 37°C for 1 h.

10 Add 10µl of unlabelled methionine (8 mg ml^{-1}), and incubate for a further 5 min.

11 Harvest the cells (3 min, Eppendorf centrifuge) and discard the supernatant.

12 Resuspend the cells in 40µl buffer B and add an equal volume of SDS-sample buffer. Heat for 3 min in a boiling waterbath.

13 Load 40 µl onto a SDS-polyacrylamide gel (Appendix C.II). The rest of the sample may be stored at −20°C.

Trouble shooting

1 If the post-UV-incorporation of label into the control strain lacking a plasmid is rather high, it may be necessary to increase the UV dose. Plating for survivors (which should be fewer than 20 viable cells ml^{-1} wth CSR603) is also a good indication of the efficiency of the irradiation. When using a strain which only lacks recA-dependent repair pathways, it has been reported that a thirty-fold greater UV-dose than that used in our experiment may be necessary. This highlights the need to optimize the conditions of irradiation for the particular strain and plasmid involved. If contamination of the overnight culture with growing cells persists, ampicillin may be added (as in the experiment here) to a concentration of 30 μg ml^{-1} provided the plasmid carried does not encode β-lactamase.

2 This technique, as with mini-cells, is more reliable if the plasmid used is relatively small and present at a high copy number. However, proteins encoded by large, low copy-number plasmids have been successfully labelled in our laboratory.

References

Isberg, R.R., Lazaar, A.L. & Syvanen M. (1982) Regulation of Tn5 by the right-repeat proteins: Control at the level of the transposition reaction? *Cell*, **30**, 883.

Kacinski, B.M., Sancar, A. & Rupp, W.D. (1981) A general approach for purifying proteins encoded by cloned genes without using a functional assay: isolation of the uvrA gene product from radiolabeled maxicells. *Nucl. Acid. Res.*, **9**, 4495.

Sancar, A., Hack, A.M. & Rupp, W.D. (1979) Simple method for identification of plasmid-coded proteins. *J. Bacteriol.*, **137**, 692.

Experiment C.3 Mini-cells

Introduction

Strains of *E. coli* carrying the mutations *min*A and *min*B divide asymmetrically so that approximately one in two divisions results in the formation of a small, anucleate cell. These mini-cells may be separated from the larger, viable cells by their differential sedimentation through sucrose gradients. When mini-cell-producing strains carry small, multi-copy plasmids (e.g. pACYC184), most mini-cells will contain, by chance, at least one plasmid molecule. Since mini-cells are capable of carrying out protein synthesis for a limited period, those containing plasmid molecules will synthesize plasmid-encoded proteins under essentially *in vivo* conditions.

The mini-cell strain DS410 has been transformed with three plasmids: pACYC184, pLG517 and pBS42 (see Table C.2), and in this experiment the objectives are:
1 Purify mini-cells.
2 To label the plasmid-encoded proteins, and to identify the dacA, dacC, Cm^R, Km^R and Tc^R gene products.

Materials needed

Nutrient broth
Streptomycin sulphate (100 mg ml^{-1})
Tetracycline hydrochloride (10 mg ml^{-1})
Kanamycin sulphate (10 mg ml^{-1})
Sodium ampicillin (1·5 mg ml^{-1})
Minimal medium
Minimal medium containing 20% sucrose
Minimal medium containing 30% glycerol
Difco methionine assay medium (10·5%)
Methionine (8 mg ml^{-1})
^{35}S-methionine
Buffer B
SDS-sample buffer
Bacterial strains
  DS410
  DS410 [pACYC184]
  DS410 [pLG517]
  DS410 [pBS42]

Procedure

1 Set up four cultures in NB (400 ml) according to the schedule shown in Table C.5 and inoculate with a single colony of each mini-cell-producing strain.

Table C.5

Strain	Plasmid	Antibiotic to be added
DS410	—	str*
DS410	pACYC184	str, tet†
DS410	pLG517	str, tet
DS410	pBS42	str, kan‡

*0·4 ml; †0·6 ml; ‡1·0 ml.

2 Place the cultures in an orbital shaker at 37°C, and allow to grow to a thick culture overnight.

3 Prepare continuous sucrose gradients by the freeze-thaw method. Prepare 500 ml of minimal medium containing 20% sucrose and:

(a) Pipette 35 ml into clear 50 ml centrifuge tubes and cover with Nescofilm.

(b) Place in −80°C freezer for 1–2 h until completely frozen.

(c) Remove gradients to the cold room, and allow to thaw out overnight. This freeze-thaw results in the formation of a perfectly adequate, reproducible, continuous sucrose gradient.

4 Chill cells at 4°C for 10 min.

5 Centrifuge in GS3 bottles (2,000 rpm, 5 min) in order to remove some of the whole cells. Pour the supernatant into fresh bottles.

6 Harvest the mini-cells and remaining whole cells (8,000 rpm, 15 min). Discard the supernatant.

7 Resuspend the pellet in 6 ml minimal medium, and layer onto two sucrose gradients, balancing them against each other.

8 Centrifuge (5,000 rpm, 18 min) in a swing-out rotor.

9 Remove the top two-thirds of the mini-cell band from each tube with a sterile plastic Pasteur pipette, and pool them.

10 Add an equal volume of minimal medium and harvest the mini-cells (13,000 rpm, 10 min).

11 Resuspend in 3 ml minimal medium, and layer onto one sucrose gradient.

12 Centrifuge (5,000 rpm, 18 min) in a swing-out rotor. Again, remove the top two-thirds of the mini-cell band.

13 Add an equal volume of minimal medium, or enough to increase the volume to at least 6 ml.

14 Measure the suspension into a clean test-tube, and note

the volume. Read the optical density at A_{600}, using distilled water as a blank.

15 Transfer the mini-cells to a centrifuge tube and harvest (13,000 rpm, 10 min). Examine under a microscope for whole cell contamination.

16 Resuspend the mini-cells in minimal medium to A_{600} 2·0.

17 Remove 100 μl of each mini-cell preparation to a sterile Eppendorf tube, and add 2 μl of a freshly prepared, sterile-filtered, sodium ampicillin solution (1·5 mg ml^{-1}). Retain the remainder for step 24.

18 Incubate the mini-cells at 37°C for 1 h, to allow long-lived mRNA to be degraded.

19 Add 2 μl ^{35}S-methionine (approximately 25 μCi), and 3 μl Difco methionine assay medium, and incubate for 30–60 min at 37°C.

20 Add 10 μl prewarmed methionine (8 mg ml^{-1}), and incubate for a further 5 min, in order to complete synthesis of partially-labelled peptides.

21 Harvest the mini-cells (Eppendorf centrifuge, 5 min).

22 Resuspend the pellet in 60 μl buffer B and add 60 μl SDS-sample buffer.

23 Heat the sample for 3 min in a boiling waterbath, and load 30 μl onto a SDS-polyacrylamide gel. The remainder of the sample may be stored at −20°C.

24 Harvest remaining mini-cells (step 17) (Eppendorf centrifuge, 2 min), discard the supernatant and resuspend in minimal medium containing 30% glycerol, again to A_{600} 2·0. Keep at −20°C. The unlabelled mini-cells may be stored for several months.

Trouble shooting

1 A relatively poor separation of mini-cells from whole cells may sometimes occur, resulting in high background incorporation of label into whole cells which can obscure the final analysis. We find in particular that cloning certain genes coding for envelope proteins in *E.coli* onto multicopy plasmids can result in disturbed division patterns in DS410, with consequent poor yields of purified mini-cells. In some cases, however, reduction of the background incorporation into whole cells can be achieved by the inclusion of ampicillin during the 60 min preincubation period, as in the experiment described. If the plasmid carried by DS410 encodes β-lactamase, then 100 μg ml^{-1} cycloserine may be used instead.

2 The Difco methionine assay medium is an amino acid mixture which lacks methionine. While its presence is not essential, we find that it enhances incorporation.

3 Whole cell contamination of mini-cells can be estimated in three ways:

(a) Under the microscope—1 whole cell per field, using mini-cells at A_{600} 2·0 and 400× magnification, is an acceptable level of contamination.

(b) Plate onto nutrient agar—10^5 viable cells ml^{-1} is an acceptable level of contamination.

(c) Label mini-cells and run on an SDS-polyacrylamide gel followed by autoradiography.

References

Reeve, J.N. (1977) Bacteriophage infection of minicells. *Mol. Gen. Genet.*, **158**, 73.

Reeve, J.N. (1979) Use of minicells for bacteriophage-directed polypeptide synthesis: *Meth. Enzym.*, **68**, 493.

Experiment C.4 Polypeptides synthesized from DNA templates *in vitro*

Introduction

By the procedure described in Appendix C.IV a cell-free extract can be prepared from *E. coli* strains, which contains the enzymes and initiation factors necessary for both transcription and translation. By supplementing this extract with nucleotide triphosphates, magnesium, amino acids and an energy-regenerating system (low molecular weight mix, LMM) and priming the system with phage, plasmid or linear DNA, one can obtain synthesis of polypeptides specified by the added template. The ability to programme polypeptide synthesis with linear DNA confers a tremendous advantage on this system since restriction enzyme fragments can be cut from large molecules and used as templates. This facilitates the mapping of polypeptide products to relatively small, well-defined regions of either the vector or the cloned DNA.

Background

Two S30 extracts (prepared as described in Appendix C.IV) will be used in this experiment. One has been produced from the strain MRE600, which lacks the major RNase activity. Since the synthesis of proteins from DNA templates relies on the production of mRNA as an intermediate, one would predict that an extract high in RNase activity would not be very efficient in protein synthesis. This is evident if DNA used to prime the system is contaminated with RNase (often used during the preparation of DNA), in which case protein synthesis is completely abolished. However, since active S30 extracts have been successfully prepared from RNase+ strains, it appears that endogenous RNase activity in *E. coli* must be relatively low, at least under the reaction conditions used. Extracts prepared from MRE600 do, however, have high DNase and exonuclease V activity and this leads to reduced efficiency when programming the system with linear DNA due to rapid digestion of the template. This problem can be overcome by using a large quantity of linear DNA. However, this is not always available. An alternative approach is to use a strain, N138, carrying a mutation in one of the genes encoding the exonuclease enzyme, (the *rec*B gene) as the source of S30.

As described in Appendix C.IV, during its preparation the S30 extract is incubated in the presence of ATP prior to final dialysis and storage (see section g, Appendix C.IV). This incubation is necessary to allow the extract to degrade any long-lived mRNA. Incubation also allows digestion of remaining chromosomal DNA fragments which were too small to be pelleted during the high speed centrifugation step. If this incubation is not carried out, the extract will support high levels of protein synthesis in the absence of added DNA, leading to an unacceptably high background. From this observation it became obvious that a functional exonuclease enzyme (*recBC* gene product), which is ATP dependent, was necessary during this incubation step. Therefore a temperature sensitive *recB* mutant is required if successful S30 extracts are to be obtained. The preparation of the extract in this case is carried out at 30°C, the permissive temperature to allow degradation of endogenous DNA. In subsequent experimental reactions with linear DNA templates at 37°C, the non-permissive temperature, the recBC enzyme is inactive and linear templates can survive for at least 2 h.

The objectives of this experiment are:

1 To identify the gene products of the *bla*, Cm^R, Tet^R, and *dac*C genes.
2 To compare the synthesis of proteins from supercoiled and linear templates using the two S30 extracts provided.
3 To show that certain restriction sites cut within coding sequences.
4 To show whether *dac*C has been cloned with its own promoter, or is using a promoter on the vector.
5 To compare the products synthesized *in vitro* with those produced from the same plasmids in maxi- or mini-cells (essentially *in vivo* systems).

Materials needed

DNA provided as follows (all 250 µg ml^{-1}):
  pBR325
  pBR325×*Eco*RI
  pBR325×*Hind*III
  pBR325×*Pst*I
  pACYC184
  pACYC184×*Eco*RI
  pLG517
  pLG517×*Eco*RI
TE buffer (10 mM Tris; 1 mM EDTA, pH 7·5)
LMM low molecular weight mix
0·1 M Mg acetate

^{35}S-methionine SJ.204
S30 extracts*
  MRE600
  N138 recB^{ts}
Methionine (8 mg ml^{-1})
SDS sample buffer

Procedure

1 Digest the three plasmid DNAs pBR325, pACY184 and pLG517 (Table C.2) with restriction endonucleases (*Eco*RI, *Hind*III and *Pst*I) as described in Section B. Ethanol precipitate the products of digestion and resuspend in TE buffer to the starting concentration.

2 Wearing gloves and using sterile technique to avoid the introduction of nucleases, prepare sterile Eppendorf vials in duplicate containing the constituents described in Table C.6.

3 Spin very briefly in an Eppendorf centrifuge to ensure that all additions have mixed.

4 Place tubes in a rack in a 37°C waterbath for 3 min.

Table C.6 Reaction mixtures to be set up for *in vitro* transcription/translation

DNA	Vol	LMM	Mg (μl)	^{35}S-methionine	TE
1 pBR325	10	7.5	3.5	2	2
2 pBR×EcoRI	10	7.5	3.5	2	2
3 pBR×HindIII	10	7.5	3.5	2	2
4 pBR×Pst 1	10	7.5	3.5	2	2
5 —	—	7.5	3.5	2	10
6 pACYC184	10	7.5	3.5	2	2
7 pACYC184×EcoRI	10	7.5	3.5	2	2
8 pLG517	10	7.5	3.5	2	2
9 pLG517×EcoRI	10	7.5	3.5	2	2
10 —	—	7.5	3.5	2	12

5 To one vial of each pair add 5 μl of S30 extract MRE600 and to the other add 5 μl of S30 N138 extract. Mix gently with the pipette tip before replacing in the waterbath.

6 After 30 min incubation at 37°C, add 10 μl of prewarmed methionine (8 mg ml^{-1}) and incubate a further 5 min.

7 Place incubations on ice and add 30 μl of buffer B, mix well.

8 Add 30 μl of SDS sample buffer and heat immediately for 3 min in a boiling waterbath.

9 Analyse 30 μl by SDS-PAGE and store the remainder at −20°C.

---

*The S30 extract from MRE600 and the LMM mix are available as a kit from Amersham International.

Trouble shooting (see also Appendix C.IV)

1 The DNA used to prime the system must be RNase-free and preferably purified from a CsCl gradient. DNA which is contaminated in some way (with ribonucleases or high salt for example) will inhibit incorporation of ^{35}S-methionine into TCA precipitable counts to such an extent as to give a value lower than the control (minus DNA). Phenol extraction and/or dialysis may overcome these problems.

2 Fragments isolated from a gel must be free of agarose.

3 The magnesium optimum estimated to produce maximum protein synthesis may not be optimal for an individual promoter and its protein product.

4 Usually ^{35}S-methionine is used as the amino acid label, alternative amino acids should be used to ensure that all proteins synthesized are being identified.

References

Collins, J. (1979) Cell-free synthesis of proteins coding for mobilisation functions of ColEI and transposition functions of Tn3. *Gene*, **6**, 29.

Lindhal, L., Post, L. & Nomura, M. (1976) DNA Dependent *in vitro* synthesis of ribosomal proteins. *Cell*, **9**, 438.

Pratt, J.M., Boulnois, G.J., Darby, V., Orr, E., Wahle, E. & Holland, I.B. (1981) Identification of gene products programmed by restriction endonuclease DNA fragments using an *E. coli in vitro* system. *Nucl. Acid. Res.*, **9**, 4459.

Yang, H., Heller, K., Gellert, M. & Zubay, G. (1979) Differential sensitivity of gene expression *in vitro* to inhibitors of DNA gyrase. *Proc. natl. Acad. Sci. U.S.A.*, **76**, 3304.

Yang, H., Ivashkiv, L., Chen, H.Z., Zubay, G. & Cashel, M. (1980) Cell-free coupled transcription-translation system for investigation of linear DNA segments. *Proc. natl. Acad. Sci. U.S.A.*, **77**, 7029.

Zubay, G. (1973) Cell-free studies on the regulation of the lac operon. *Ann. Rev. Gen.*, **7**, 267.

Appendix C.I Some general media used in Section C

Nutrient broth: 2·5% nutrient broth: nutrient broth powder

M9-salts
   Na_2HPO_4 0·4 M
   KH_2PO_4 0·2 M
   NaCl 80 mM
   NH_4Cl 0·2 M
   Ca/Mg salts
   $CaCl_2$ 0·1 M; $MgSO_4$ 1M
   Glucose (or maltose) 20% w/v
   Amino acids as required 2 mg ml^{-1}

Buffer B
   Na_2HPO_4 7 g l^{-1}
   KH_2PO_4 3 g l^{-1}
   NaCl 4 g l^{-1}
   $MgSO_4$ 0·1 g l^{-1}

Lambda buffer (LB)
   Tris 6 mM (pH 7·2)
   $MgSO_4$ 10 mM
   Gelatin 0·5 g l^{-1}

K medium
   M9-minimal medium + 0·2% w/v Difco casamino acids
                                  + 0·1% w/v thiamine

Hershey salts (sulphur free)
   NaCl 5·4 g ⎫
   KCl 3·0 g   ⎪
   NH_4Cl 0·087 g ⎬ autoclave
   Tris 12·1 g ⎪
   HCl to pH 7·4 ⎪
   H_2O to 1 litre ⎭
   To each 100 ml of above add:
   $CaCl_2$ $2H_2O$ 1 M 0·1 ml
   $MgCl_2$ $6H_2O$ 1 M 0·1 ml
   $FeCl_3$ $6H_2O$ 1 mM 0·1 ml

Hershey medium
   To 100 ml of complete Hershey salts add:
   2 ml 20% glucose

1 ml amino acid requirements:
CSR603 thr 10 mg ml^{-1}
leu 10 mg ml^{-1}
pro 20 mg ml^{-1}
arg 20 mg ml^{-1}
thi 0·1 mg ml^{-1}

CSH26 is *pro thi*, so the same supplement is used.

Appendix C.II SDS-polyacrylamide gel electrophoresis

Background

The results of experiments involving the analysis of polypeptides by electrophoresis can frequently be rendered inconclusive if such analyses are not carried out with extreme care and good quality materials. Several different methods are available but the one we have found most successful is essentially that described by Laemmli (with minor modification, see below) in which polypeptides are electrophoresized in the presence of sodium dodecylsulphate (SDS).

SDS-PAGE allows proteins to be fractionated more or less according to molecular weight. The gel is formed by polymerization of acrylamide and N,N'-methylene-bisacrylamide [catalyzed by ammonium persulphate and N,N,N'-Tetramethyethylenediamine (TEMED)]. The degree of cross-linking in the gel is governed by the ratio of acrylamide to bisacrylamide used. We generally use a ratio of 44:0.8, but this is not optimal for all protein separations, and may be varied.

Proteins are boiled in SDS-sample buffer before being loaded onto the gel. SDS in the buffer binds to the hydrophobic regions of proteins, leading to their solubilization, and rendering them more or less uniformly negatively charged. They may then be electrophoresed towards the anode (+), and their mobility depends mainly on the molecular weight of the protein and the concentration of the acrylamide gel. Electrophoresis grade SDS generally gives more reliable results, and is included in the gel and the electrophoresis buffer.

A discontinuous gel system is used. A low percentage stacking gel is poured on top of the main separating gel. The stacking gel is prepared at a lower pH than the separating gel. As the proteins pass from one gel to another, there is a 'stacking' effect, so that sharp bands are produced however much sample is loaded in the slots.

After electrophoresis, the proteins in the gel may be stained using Coomassie brilliant blue, but since the samples described here are all radioactively-labelled, this is not necessary. The proteins in the gel are 'fixed' by immersion in 10% acetic acid, 25% propan-2-ol for a short period, and the gel is then fluorographed (see Appendix C.III). This involves impregnation of the gel with the scintillant 2,5-diphenyloxazole (PPO), having replaced all water in the gel with DMSO. The gel is then soaked in several changes of water to precipitate the PPO and remove the DMSO. The gel is dried down under vacuum on to thick card, and exposed to X-ray film at $-80°C$. If the incorporation of label is high enough, the gel need not be fluorographed. In this case, after fixing the proteins, the gel is dried down, and exposed to X-ray film at room temperature.

Preparation of gels

Working from the gel solutions indicated in Table C.5 (and see Tables C.4, C.6) gels are cast and operated as follows:

Procedure

Assembly of gel kit

1 Clean the glass plates with dichromate cleaning solution, concentrated HNO_3 or acetone. Dry the plates completely and avoid touching the glass plate surfaces. We have found Pyroneg adequate, with chromic acid every 4th or 5th run.

2 Place the centre core of the cell on its side. Place a notched glass plate on the flat acrylic surface and align the glass plate notch with the acrylic plate notch.

Table C.7 PAGE equipment and chemicals. Vertical slab gel unit, model 220, Biorad

Chemical	Supplier
Acrylamide	Biorad (expensive but pure)
Bis acrylamide	Eastman (Kodak) cheaper but clean with charcoal before use
SDS*	Biorad (must be very pure) 161 0300
TEMED	Biorad 161 0800
Ammonium persulphate	Biorad 161 0700
β-mercaptoethanol	Sigma M 6250
Tris	Sigma T 1503
Coomassie Blue	Biorad 161 0400

*SDS precipitates at 4°C. Solutions kept in the fridge thus need warming to room temperature to dissolve SDS before use. SDS also precipitates in the presence of K^+, so good distilled water is needed, and the use of K^+ salts should be avoided in the preparation of proteins for gel analysis.

3 Lightly grease and align the PVC sandwich spacers along the two outer edges of the glass plate, taking care that they are exactly level with the edge of the glass plate at the bottom end.

4 Place a rectangular glass plate over the spacers to form the sandwich. Check again that the spacer ends and the bottom edges of the glass plates are exactly aligned.

Table C.8 Acrylamide reagents

(a) Separating gel buffer
  0·75 M Tris-HCl, 0·2% SDS w/v, pH 8·8

(b) Spacer gel buffer
  0·35 M Tris-HCl, 0·2% SDS w/v, pH 6·8

a and b: Dissolve Tris and SDS; adjust pH with HCl; make up to volume.

(c) Sample buffer
  2 ml glycerol
  0·4 g SDS
  5 ml 0·25 M Tris-HCl, pH 6·8
  2 ml H_2O
  1 ml β-mercaptoethanol store refrigerated

(d) Acrylamide	DI	DII	DIII	DIV
Acrylamide monomer	44	44	30	30
Bis dimer	0·8	0·3	0·3	0·8

Note 1. To purify acrylamide—dissolve in water, mix with activated charcoal, stir 10 min, filter through course and then fine filter paper (Whatman No. 1), store refrigerated, protect from light.

Note 2. Resolution in particular size ranges or for specific proteins can be markedly affected by varying the monomer:dimer ratio of acrylamide. DIV is the original Laemmli recipe and is particularly poor if proteins in the 75–90 K region are to be resolved. DI is much preferred for this and is a good all purpose recipe. DII is particularly good for resolving in the 30–40 K region and DIII is essential for good resolution of penicillin-binding proteins.

Note 3. Different % acrylamide gels are used according to the size of the peptides to be analysed, see Table C.9. For an initial screening of products 11% is best. This resolves proteins within the range 25–180 K reasonably well. The best recipe for peptides 60–240 K is 5%, for 50–160 K is 7·5% and 25–100 K is 11%, for 10–60 K is 15%, < 25 K is 20%.

Laemmli, U.K. (1970) *Nature*, **227**, 680.
Hancock, R.E.W., Hankte, K. & Braun, V. (1976) *J. Bacteriol.*, **127**, 1370.

---

5 Slip two clamps on each side of the unit (four clamps for the Model 221). Use two (four) clamps along each edge. Before tightening the clamps, check again as in stage 4. Test the comb to see if it inserts easily. If not, loosen clamps for easier comb insertion. Lightly grease the bottom ends of the spacers and the edges of the plates they lie against.

6 Place a bottom support-sealing bar at the bottom of the glass plates. Insert two screws through the slots of the bar into the threaded holes on the centre core. Do not screw tight.

7 Push the support sealing bar with the sealing gasket up to seal the lower sandwich space. Insert the cams and turn to apply upward pressure to seat the support bar against the lower edge of the glass plates. Secure the sealing bar in place by gently tightening the screws.

8 Turn the cell over and repeat steps 2 through 7 for the second sandwich. If only one slab is made, clamp on a reservoir plate to create the upper well and position the second support sealing bar on the core to form a set of legs.

Filling the slab sandwich

1 Place the cell on the levelling plate and position the L-shaped bubble level on the centre of one of the outer glass surface plates with the long leg of the level flush against the outer glass surface. Align the bubble by adjusting the levelling plate screws.

2 The gel solution can be fed from the bottom of the sandwich, using a syringe needle attached by tubing to a gradient marker or gel reservoir or syringe. For non-gradient gels we find it perfectly satisfactory simply to pipette in from the top. Pump or pipette acrylamide solution into the sandwich to a level approximately 1 cm below the bottom of the well-forming comb, i.e. about 3·5 cm below the edge of the notch. This distance represents the height of the stacking gel and may be varied at discretion (a flow rate of 0·5–1·0 ml min^{-1} is recommended for pumping of acrylamide gradients).

Table C.9 Gel composition.

	Acrylamide concentration		
Separating gel	8·5%	11%	15%
Buffer 'A'	13·5	13·5	13·5
Acrylamide solution DI 'C'	5·3	6·8	9·2
Distilled water	7·5	6·0	3·6
Ammonium persulphate 10 mg ml^{-1}	0·95	0·95	0·95
TEMED	75 μl	75 μl	75 μl

Volumes are enough for 1 gel and concentration depends on whether low molecular weight proteins (15%) or larger proteins (8·5%) are being analysed

Mix 'A', 'C' and distilled water; de-aerate, by attachment of flask to vacuum line, keeping solution cold. Then add ammonium persulphate, freshly dissolved, and TEMED. Now work quickly to make the gel as polymerization will start in 10–15' as solutions warm up. Take care not to produce bubbles as these will be stable and difficult to remove. When the gel solution is in place spray 0·1% SDS on to the surface

Stacking gel: 7% acrylamide
  Buffer 'B'   10 ml
  Acrylamide solution 'C'   3·3 ml
  Distilled water   6·7 ml
  Ammonium persulphate 10 mg ml^{-1} 0·5 ml (freshly made)
  TEMED   40 μl
Enough for 2 gels

Staining solution
  0·05% w/v Coomassie blue dissolved in 10% v/v acetic acid and 25% v/v isopropanol
  300 ml stain for one gel. Shake very gently overnight

Destaining procedure
  Pour off stain taking care not to let the gel fold and crack.
  Pour on 10% acetic acid, 10% isopropanol, shake, change destain frequently until background is clear

3 Cooling water now may be connected to the inner core if necessary. For short runs this does not appear to be necessary.

4 Spray the plate above the gel with 0·1% SDS to overlayer the gel for a smooth surface after polymerization. Use an ASL-airflow 'Spraymist' for this. If necessary gels can be kept overnight in the cold at this stage.

5 After polymerizing the separating gel, carefully invert the unit to drain the overlayering fluid from the sandwich. Rinse the acrylamide surface twice with separating gel buffer or distilled water and drain excess fluid. All of the fluid can be removed by inserting a piece of filter paper into the sandwich space and absorbing all drops along the width of the sandwich. Do not damage the gel surface.

6 With the cell returned to an upright position, dry and reinsert the desired sample well-forming comb. Introduce the stacking gel at the side of the comb with a Pasteur pipette.

7 After polymerizing the stacking gel, pour some Tris-glycine buffer into the upper reservoir, gently remove the comb and rinse the sample slots twice with the buffer using a syringe or pasteur pipette. Do not put bubbles into the slots.

8 Remove the support-sealing bars from the cell.

9 Place the electrophoresis unit into the lower buffer box. Fill the lower buffer chamber and remove any bubbles on the lower gel surface by tilting the electrophoresis unit. Put the lower buffer chamber box on the levelling plate and use the L-shaped bubble level in the same manner as described in step 3 to ensure that the slab sandwich plate is level.

10 Mix the sample, 5–60 μl containing up to maximum of 100 μg protein with SDS-sample buffer to make the sample more dense than the buffer solution. A 10% concentration of glycerol is normally sufficient. Minute quantities of tracking dye (bromophenol blue in 25% glycerol) may be added to the sample to observe the rate of migration. Heat the samples for 3 min in a boiling waterbath and apply them to the slots through the buffer, introducing the syringe needle 2/3 down the slot.

11 Place the lid on the buffer box. Optionally attach the coolant fitting. Be sure they do not leak.

12 Connect the electrode leads to a suitable D.C. Power Supply and electrophorese 100–150 V at constant volts, 15 mA gel^{-1} applied until the dye front is well into the stacking gel (about 1 h). Then 25 mA gel^{-1} at constant current applied for 3–4 h or until the dye front is close to the bottom of the gel.

Gel removal

1 The glass plates are separated after the run by removing the spacers and inserting a straight edge along one side of the sandwich. Apply gentle pressure to pry the plates apart. If the plates do not readily separate, remove the glass plate by gently directing a stream of water from a fine gauge needle (22 gauge) between the acrylamide slab and the glass.

2 To remove the gel from the remaining plate, hold the plate upside down, with the gel underneath, over a bath of stain, just few inches from the liquid. Loosen a corner of the gel with a knife or spatula, and the gel should gently drop into the stain without folding. The gel is fragile and brittle and cannot easily be picked up and handled without breaking.

Appendix C.III Autoradiography and fluorography of gels

Gels containing less than 10^5 cpm of, for example, ^{35}S-methionine per sample can be treated for fluorography. Impregnation of gels with scintillant before drying and exposure to X-ray film results in enhancement of soft β-emission by factors of around 15-fold. Two methods are commonly in use, one using impregnation of gels with PPO (diphenyloxazole), the other with sodium salicylate (or a related commercial preparation).

PPO method

Procedure

1 After fixing the gel, place it in approximately 250 ml dimethyl-sulphoxide (DMSO) for 30 min with gentle shaking (this dehydrates the gel, in preparation for impregnation with H_2O-insoluble PPO). Remove DMSO.

2 Gently shake again for 30 min in the same volume of fresh DMSO. Remove and retain DMSO (may be used again).

3 Shake gel in 20% PPO dissolved in DMSO for about 2 h. Remove PPO solution, which can be reused several times.

4 Rehydrate and thoroughly wash the gel by treating with slowly running tap water for 30–60 min until no trace of DMSO can be discerned, alternatively the gel can simply be soaked for 60 min in water.

5 The next stage applies whether the gel has been stained, merely fixed, or impregnated with DMSO/PPO as described stages 1–4. Gels are dried on to Whatman chromotograph paper No. 17 or Whatman No. 1 (if bands are to be cut out and counted in a quantitative experiment) using a Biorad Model 224 slab gel drier connected to a KNF Laboport vacuum pump. Drying may take 1–3 h.

6 Fluorography or autoradiography is carried out by placing the dried gel in X-ray cassette with a sheet of Kodak RP Royal X-ray film, storing at room temperature for autoradiographs or at $-80°C$ for fluorographs, for times varying from 24 h for samples containing $> 10^6$ cpm, to 2 weeks for samples containing 10^4 cpm.

When gels contain low levels of radioactivity, films are

pre-fogged by limited exposure to photographic flash to increase sensitivity.

7 Films are developed in darkness in Kodak DX-80 developer for 4 min followed by an acid wash (1 min in 1% acetic acid and 4 min in Kodak FX-40 fixer).

Care: DMSO is a skin irritant and may also facilitate uptake of carcinogenic PPO. Use gloves when handling these compounds.

References

Bonner, W.M. & Laskey, R.A. (1974) A film detection method for ^{3}H-labelled proteins and nucleic acids in P.A.G. *Eur. J. Biochem.*, **46**, 83.
Chamberlian, J.P. (1979) *Anal. Biochem.*, **98**, 132.
Laskey, R.A. & Mills, A.D. (1975) Quantitative film detection of [^{3}H] and [^{14}C]- in P.A.G. by fluorography. *Eur. J. Biochem.*, **56**, 335.

Use of fast tungstate screens in autoradiography

Hard β-emission (such as from ^{32}P) or γ-emission from ^{125}I which passes through the X-ray film can be intensified by use of tungstate intensifying screens.

Procedure

1 Cover a wet gel with Clingfilm; dried gels can be exposed directly.
2 Make a sandwich of: Gel sample; X-ray film; tungstate screen (Fugi Mach 2); white side next to film.
3 Insert in light-proof case, place under heavy weight.
4 Store $-70°$C.

Reference

Laskey, R.A. (1980) *Meths. Enzymol.*, **65**, 363.

Developing autoradiographs

Material needed

Kodak developer D-19
Kodak fixer 'Amfix'
H_2O rinse

Procedure

1 Set up above solutions in trays in darkroom. Red safety light can be used.

2 Retrieve gel sample from $-70°C$ freezer.

3 Immediately go to darkroom (i.e. do not let H_2O vapour form on film, which may cause sticking to gel), and, working with a red safety light, place film in developer for up to 5 min, or until bands or spots are strong enough. Rinse in H_2O, then fix until film 'clears'.

4 Wash extensively in running tap water.

5 Hang up to dry.

Appendix C.IV Preparation of the components of an *E. coli in vitro* transcription/translation system

Treatment of water for preparation of all reagents and S30 buffer

One millilitre of diethylpyrocarbonate (DEPC) per litre of distilled water is added slowly (with stirring). Stirring is continued for 1 h and the mixture is then autoclaved and stored at 4°C. For each preparation 3×4 l and 8×500 ml is usually needed, i.e. a total of 16 l.

Treatment of apparatus

All glassware and plastic tubing is treated with DEPC by filling (or soaking) for 1 h with water to which DEPC has been recently added. The apparatus is then drained and autoclaved. A typical requirement is as follows:

6×1 litre flasks with screw top
$16 \times$ SS34 tubes + caps
6×25 ml beakers
4×100 ml or 50 ml measuring cylinders
$6-10 \times$ McCartney bottles
$3-4 \times$ stirring bars
$20-30 \times$ Pasteur pipettes
3×100 ml Büchner flasks
$1 \times$ Bung for above
Several spatulas and teflon homogenizing pestle
Dialysis tubing (eight pieces of 1″ diameter, 12″ long), which has been boiled in 0.1 M $Na_2CO_3/0.01$ M EDTA solution, rinsed extensively in distilled water and then stored in sterile DEPC treated water at 4°C

Adjustment of pH

The pH of alkaline solutions is adjusted with acetic acid. For those which are too acid use 2 M Tris (in DEPC-treated water). In some cases 1/10th acetic acid (1 ml glacial acetic + 9 ml DEPC water) is required (e.g. to adjust the pH of the PEP). Both solutions should be autoclaved before use.

The stock solutions necessary for the preincubation mix and the low molecular weight mix are then prepared as outlined in Table C.10 and C.11.

Table C.10 Preparation of solutions. The following solutions for the low molecular weight mix and preincubation mix are prepared in DEPC-treated water

	Solution	Weight	Volume (ml)	Aliquots
i	2·2 M Tris acetate, pH 8·2	26·65 g	100	
ii	3 M Mg acetate, 4 H$_2$O	64·2	100	
iii	0·55 M DTT	424 mg	5	130 µl, 25 µl
iv	38 mM ATP (pH 7·0 with Tris)	419 mg	20	2·7 ml, 220 µl
v	88 mM CTP	42 mg		
	GTP	46 mg	1	70 µl
	UTP	42 mg		
	(pH 7·0 with Tris)			
vi	0·42 M PEP	938 mg	10	1·6 ml, 420 µl
vii	40% polyethylene glycol (6,000)	4 g	10	
viii	Folinic acid (Ca^{2+} leucovorin)	2·7 mg	1	100 µl
ix	50 mM cAMP, pH 7·0	16 mg	1	100 µl
x	Transfer RNA	17·4 mg	1	70 µl
xi	Inorganic mix			
	1·4 M NH$_4$ acetate anhydrous	1·08 g		
	2·8 M K acetate anhydrous	2·74 g	10	
	0·38 M Ca acetate anhydrous	0·6 g		
xii	0·1 M Mg acetate, 4 H$_2$O	0·21 g	10	
xiii	Methionine	8 mg	1	
xiv	Amino acid mix			
	a 50 mM + methinone			
	b 55 mM – methionine			

Amino acid	a 50 mM	b 55 mM
Alanine	22·3 mg	24·53 mg
Arginine, HCl	52·7 mg	58·0 mg
Asparagine	33·0 mg	36·3 mg
Aspartic acid	33·3 mg	36·6 mg
Cysteine	30·3 mg	33·33 mg
Glutamic acid	36·8 mg	40·5 mg
Glutamine	36·5 mg	40·15 mg
Glycine	18·8 mg	20·68 mg
Histidine, HCl	47·9 mg	52·7 mg
Isoleucine	32·8 mg	36·1 mg
Leucine	32·8 mg	36·1 mg
Lysine	45·7 mg	50·3 mg
Methionine	37·3 mg	—
Phenylalanine	31·3 mg	34·4 mg
Proline	28·8 mg	31·7 mg
Serine	26·3 mg	29·0 mg
Threonine	29·8 mg	32·8 mg
Tryptophan	51·1 mg	56·2 mg
Tyrosine	45·3 mg	49·8 mg
Valine	29·3 mg	32·2 mg

Each mixture of amino acid is added to 5 ml of DEPC-treated water and the suspensions frozen at −20°C in 50 µl aliquots.

S30 buffer

The buffer needed to wash the cells and prepare the S30 extract is made as follows:

(a) 500 ml 1 M Tris acetate, pH 8·2
(b) 500 ml 6M K acetate
(c) 500 ml 1·4 M Mg acetate
(d) 100 ml 100 mM dithiothreitol DTT

Made with DEPC water and autoclaved

Table C.11 Low molecular weight mix (LMM)

	Stock solution	Volume (μl)	Final concentration in synthesis mix
1	Tris-acetate 2·2 M, pH 8·2	40	56·4 mM
2	Dithiothreitol 0·55 M	5	1·76 mM
3	ATP 38 mM, pH 7·0	50	1·22 mM
4	CTP GTP each 88 mM, pH 7·0 UTP	15	0·85 mM
5	Phosphoenol pyruvate 0·42 M, pH 7·0	100	27 mM
6	19 amino acids 55 mM each (suspension) (i.e. -methionine)	10	353 μM
7	Polyethylene glycol-6,000 40% in H_2O	75	1·92%
8	Folinic acid 2·7 mg ml^{-1} (Ca leucovorin)	20	34·6 μg ml^{-1}
9	cAMP 50 mM, pH 7·0	20	641·0 μM
10	tRNA *E. coli* 17·4 mg ml^{-1}	15	167·3 μg ml^{-1}
11	Inorganic mix Ammonium acetate 1·4 M Potassium acetate 2·8 M Calcium acetate 0·38 M	40	36·0 mM 72·0 mM 9·7 mM

The components are mixed in the order given.

100 ml of (a), (b) and (c) are made up to 1 litre (in a pretreated flask, with DEPC-treated H_2O. This is kept at 4°C. This is a ×10 S30 buffer solution (e) which is diluted just prior to use by taking 100 ml of the latter (e) and 10 ml of solution (d) (DTT) and making up to one litre with DEPC H_2O. This will produce an S30 buffer with composition as follows:
10 mM Tris acetate, pH 8·2
14 mM Mg acetate
60 mM K acetate
1 mM DTT

Preparation of cells

Cells (MRE 600, N138 *rec*B^{ts} and K12) are grown in a fermentor in medium made up as follows:
(a)   56 g KH_2PO_4
289 g K_2HPO_4
10 g yeast extract
10–15 mg thiamine
0·5 g each required amino acid
10 litres water
Autoclaved for 20 min
(b)   Add 800 ml 25% glucose autoclaved for 20 min.
(c)   Add 100 ml 100 mM Mg acetate (autoclaved for 20 min).
(d)   Inoculate with an overnight culture (100 ml) of the appropriate strain, to give an A_{450} of approximately 0·07. The

culture is allowed to grow to an A_{450} of 2–3 at 37°C (or at 30°C for ts strains) and harvested at 4°C. While cells are being harvested the ×10 S30 buffer is diluted as described in stage 4. About 2 litres are needed. The cell pellets are washed three times in S30 buffer (+0·5 ml mercapteothanol l^{-1}: 250 ml 10 g^{-1} cells). Each reharvesting spin is carried out at 10,000 rpm in GS3 bottles for 20 min. The cell pastes are stored at −80°C overnight.

Preparation of the S30 extract

(a) Rinse a French press with DEPC treated water, wrap in foil and hold at 4°C. Hold cell pellets for 30 min at 4°C. Resuspend slowly (1 min) in S30 buffer (100 ml 10 g^{-1} cells+0·05 ml mercaptoethanol).

(b) Centrifuge 30 min, 10,000 rpm in Sorval GS3 rotor, then weigh cells.

(c) Resuspend carefully under vacuum in S30 buffer (63·5 ml per 50 g cells). Use a buchner flask for this, spoon cells into the flask. Add the S30 buffer and attach to a vacuum pump using a bung in the top. Constant swirling and mixing with a pestle combined with frequent evacuations results in cells being resuspended fairly anaerobically.

(d) French press at 8,400 p.s.i., collect 10 ml fractions and add 100 μl 0·1 M DTT per fraction as soon as it is collected. The preparation may still be viscous and turbid at this stage. Any whole cells remaining are removed in the subsequent spin. It is not advisable to attempt to improve the level of breakage by re-pressing as this leads to loss of activity.

(e) Centrifuge in SS34 rotor at 15,500 rpm for 30 min (35 ml tubes). At this point prepare the preincubation mix (Table C.12) and put 37°C waterbath on (30°C waterbath in the cases of temperature sensitive strains).

Table C.12 Preincubation mix.

Pyruvate kinase [10 mg ml^{-1} in $(NH_4)_2SO_4$]	25 μl
Tris acetate, pH 8·2 (2·2 M)	1·0 ml
Mg acetate (3 M)	23 μl
ATP 38 mM, pH 7·0	2·63 ml
PEP 0·42 M, pH 7·0	1·50 ml
DTT 0·55 M	60 μl
20 amino acid mix (50 mM)	6 μl
H_2O (DEPC treated)	to 7·5 ml

(f) Transfer the upper 4/5th to a small Erlenmeyer flask protected from the light. Add 75 ml preincubation mix per 25 ml supernatant.

(g) Incubate 80 min in a 37°C waterbath (or 160 min at 30°C). Prepare 4 litres of S30 buffer and place in cold room.

(h) Pour mixture into 1″-wide dialysis bags. Dialyse 4×45 min against 50 volumes of S30 buffer at 4°C.
(i) Centrifuge in SS34 rotor at 6,000 rpm for 10 min.
(j) Distribute the supernatant in 05 ml portions into 2 ml Nunc vials (cat. 3810). Store in liquid nitrogen. If the liquid nitrogen container is properly maintained these extracts will give reproducible activities for several years.

Making up LMM

The low molecular weight mix (LMM) may now be made up from the stock solutions described in Table C.11. This mix may be frozen and stored at −20°C for at least 2–3 months.

Optimization of the magnesium concentration

To optimize the Mg concentration set up incubations with varying volumes of a 0·1 M Mg acetate solution (Table C.13). It may be necessary to optimize the S30 concentration as well. A typical optimization experiment is shown below.

Requirements

DNA (preferably 250–500 μg ml^{-1})
^{35}S-methionine (7μCi μl^{-1})
TA buffer: 10mM Tris acetate, pH 7·0
LMM
0·1 M Mg acetate
S30 extract
9 mg ml^{-1} methionine

Table C.13 Incubations to optimize Mg concentration.

	DNA (μg)	Vol (μl)	^{35}S-methionine	LMM	Mg	TA	S30
1	1·5–2·5	5	2	7·5	2·0	8·5	5
2	1·5–2·5	5	2	7·5	2·5	8·0	5
3	1·5–2·5	5	2	7·5	3·0	7·5	5
4	1·5–2·5	5	2	7·5	3·5	7·0	5
5	1·5–2·5	5	2	7·5	4·0	6·5	5
6	1·5–2·5	5	2	7·5	2·5	5·0	8
7	1·5–2·5	5	2	7·5	3·5	4·0	8
8	—	—	2	7·5	3·5	12·0	5
9	—	—	2	7·5	3·5	9·0	8
			9·5 μl				

Incubating the mixture

The DNA, ^{35}S-methionine+LMM, TA and Mg^{2+} are mixed in a sterile Eppendorf and incubated at 37°C for 2–4 min. The S30 extract is then added to each tube and incubated with shaking for 30 min to 1 h. After this period 10 µl of 8 mg ml^{-1} methionine (prewarmed) is added and incubation carried out for a further 5 min. A 2 µl sample is then removed for TCA precipitation to estimate the level of incorporation. To the remainder, 30 µl of 10 mM sodium phosphate, pH 7·2, and 30 µl of SDS sample buffer is added, the sample is then boiled for 5 min and may be stored at $-20°C$ until analysed by SDS-PAGE. A maximum of 2–3 µl of S30 extract can be loaded on a gel before the protein will overload the slot. The TCA counts indicate the optimum Mg and S30 concentration for each extract prepared.

Care

1 Gloves and sterile equipment should be used at all times.
2 DNA must be RNase free. If RNase has been used in the course of the DNA preparation then only repeated phenol extraction will render the DNA usable in this system. Some RNA contamination is often preferable. The DNA should be free of CsCl, EtBr and high salt. Usually it is preferable to supply DNA at a concentration of 300 µg mg^{-1} ml^{-1} in 10 mM Tris, 1 mM EDTA, pH 7·0–7·5.

Linear DNA as a template

Purified restriction fragments (see Appendix B.II) from agarose gels may be used to prime the *in vitro* system. However, due to the endogenous exonuclease activity of the S30 extract, added DNA is rapidly digested and successful synthesis is only obtained at high concentrations of DNA. This problem may be overcome by preparing an S30 extract from a strain carrying a temperature sensitive mutation in the *rec*B gene, the source of the major exonuclease activity. Only minor alterations in the method are necessary for preparation of an S30 extract from a *rec*B ts strain. These are included in the detailed method described in Appendix C.IV. Using this extract as little as 50 ng of DNA is sufficient to prime the system.

DNA fragments generated by restriction endonucleases may also be analysed directly, without the need to purify specific fragments. After restriction is complete the DNA is recovered by ethanol precipitation, the pellet is dried and resuspended in TE buffer and added to the *in vitro* system.

Section D   Bacteriophage λ as a vector

Dave Burt
Bill Brammar

Introduction

The use of bacteriophage λ as a vector

Bacteriophage λ is a temperate coliphage with a double-stranded DNA genome of approximately 50 Kb. The phage DNA is readily amenable to genetic and biochemical manipulation, so that the arrangement and functioning of the viral genes is now well understood. Derivatives of phage λ have been used as cloning vectors since 1974 (Murray & Murray 1974, Rambach & Tiollais 1974, Thomas et al. 1974) and a large variety of lambdoid vectors has subsequently been constructed (for reviews see Williams & Blattner 1980, Brammar 1982, Murray 1983). Effective use of these vectors demands a basic understanding of the molecular biology of phage λ.

The arrangement of the λ genes

Although the λ genome contains about 50 genes, only 50% of these are essential for phage growth and plaque-formation. Genes of related function tend to be clustered on the genome (Fig. D.1). The essential genes concerned with head (A–F) and tail (Z–J) formation and assembly occupy the left-hand third of the genome. The region of the map between J and att, the phage attachment site, contains non-essential genes coding for proteins of unknown or unimportant function (Hendrix 1971). Genes to the right of att govern site-specific (int and xis) and generalize (redA and redB) recombination of phage DNA and the control of lysogeny (cIII). None of the genes between J and N is essential for the phage's ability to grow and make plaques, and it is largely this region of the genome that is deleted or replaced when using λ as a vector. The product of gene N is the early regulatory protein that is normally necessary to activate transcription of most other phage genes. The N gene becomes dispensable in the presence of a deletion, ninR5, that removes the transcription terminator, t_{R2}.

Gene cI is the structural gene for the λ phage repressor, the protein that switches off transcription of phage genes in the lysogenic state (Ptashne 1971). The presence of the λ repressor make a λ-lysogenic cell immune to superinfection by another λ phage and is responsible for the characteristic turbidity of λ plaques. The genes to the right of cI in Fig. D.1 are all essential for plaque-formation. The products of genes O and P

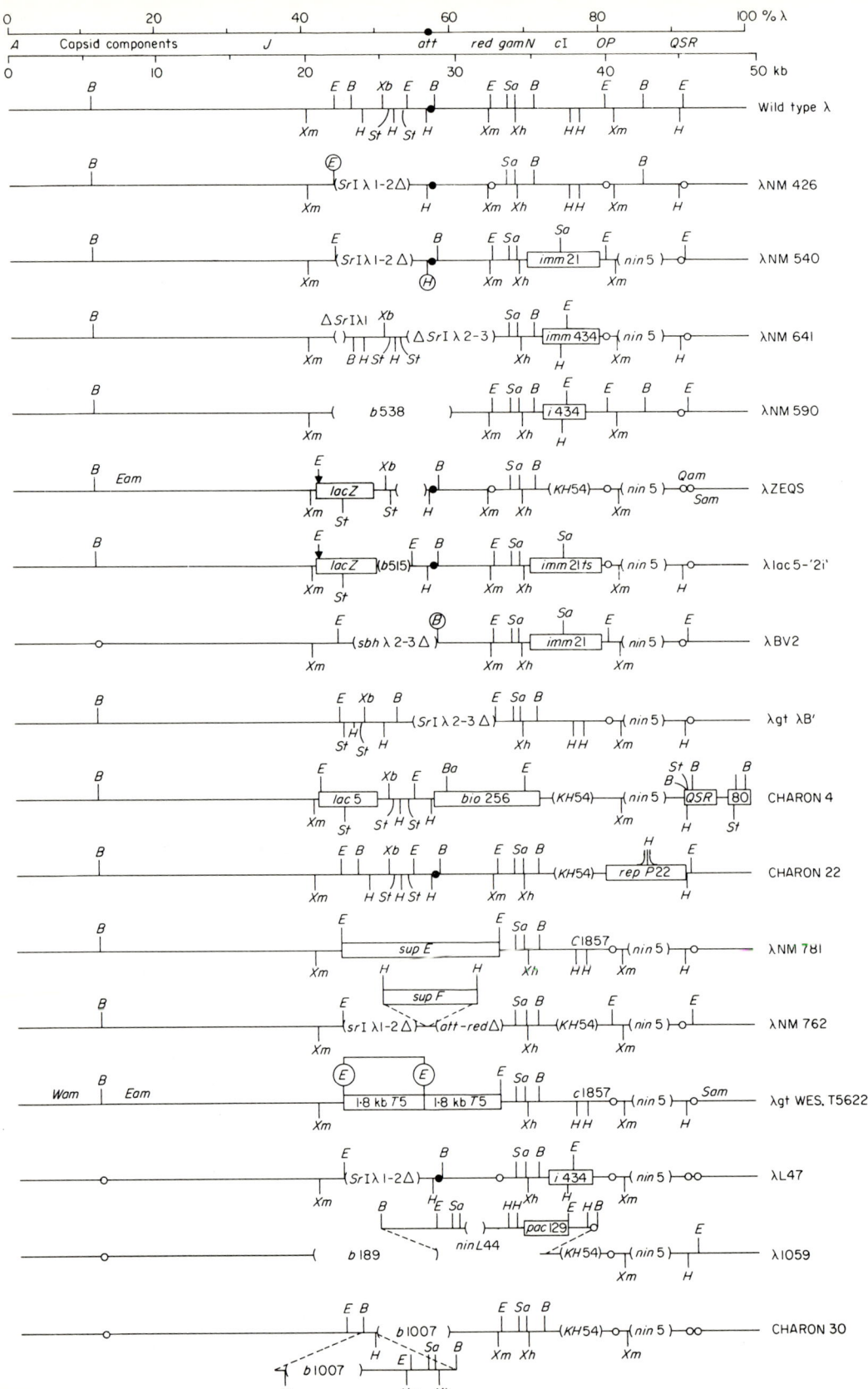

Fig. D.1 Structure of some λ vectors (see Table D.1).

are required for replication of λ DNA (Brookes 1965, Joyner et al. 1976), the Q gene product for activation of late transcription (Dove 1966, Couturier et al. 1974) and the S and R gene products for lysis of the host cells (Campbell & Del Campillo-Campbell 1963, Harris et al. 1967).

The expression of λ genes in the lytic cycle

Immediately after infection of a sensitive host or induction of a lysogenic strain, transcription by the host's RNA polymerase proceeds leftwards from the promoter p_L through gene N and rightwards from promoter p_R through the cro gene (Fig. D.2). Most of these initial transcripts terminate at sites t_{L1} and t_{R1} immediately beyond genes N and cro respectively. The termination process at t_{L1} and t_{R1} is not always effective, but the few per cent of transcripts that escape here are terminated at subsequent sites, such as t_{R2} preceding gene Q. The early leftward transcript is translated to yield the N protein, which exerts its controlling effect by influencing RNA polymerase to ignore transcription termination signals (Adhya et al. 1974, Franklin 1974, Segawa & Imamoto 1974).

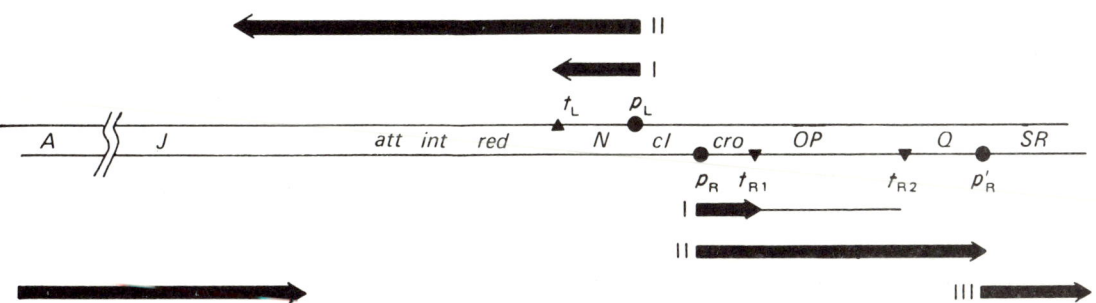

Fig. D.2 RNA polymerase transcription after infection.

In the presence of N protein, leftward transcription initiated at p_L is elongated through the gam, red, xis and int genes, while rightward transcription from p_R proceeds through the O and P genes, leading to the initiation of phage DNA replication, and through t_{R2} and gene Q. The product of the latter gene activates transcription from the late promoter, p_R', through genes S, R and the genes A to J in the circular phage chromosome.

The products of the regulatory genes N and Q are normally essential for phage development and plaque formation. Since the transcription-terminating site t_{R1} is inefficient, an N^- phage can express the O and P genes and replicate its DNA. If the terminator site t_{R2} is removed genetically, an N-defective phage can then express gene Q and proceed through late gene expression to maturity and plaque formation. Many lambdoid vectors contain a deletion of t_{R2}, $nin5$ (Court & Sato 1969) that conveniently generates extra capacity for cloned DNA within the vector and makes the phage N-independent.

The product of the *cro* gene, after a delay of about 10 min required to achieve the critical concentration, interacts directly with the template DNA at the complex p_L and p_R control sites and severely depresses the rate of the now superfluous transcription from these promoters (Radding & Shreffler 1966, Pero 1971, Echols et al. 1973).

Phage DNA replication

Bacteriophage λ DNA is replicated bidirectionally (Schnös & Inman 1970) from an origin within the O gene (Denniston-Thompson et al. 1977), in a process that requires the products of phage genes O and P acting in concert with host replication functions. During the first 10 min or so after infection the phage genomes are replicated as monomeric circular molecules, but the rolling circle mode ensues to produce the multimeric concatenates that are the substrates for packaging of linear monomeric genomes during the phage maturation process.

DNA packaging and size selection

The λ DNA packaging reaction imposes a stringent requirement for DNA size, only molecules between 78% and 105% of the wild type genome length being packaged into viable particles (Weil et al. 1973). While this property necessarily sets an upper limit on the size of DNA fragment that can be cloned in a λ vector, this limit is about 22 kb, since the essential phage genes occupy only about 28 kb of λ DNA.

Because there is both a lower and an upper limit to the size of a lambdoid phage genome, different strategies are required for the cloning of small versus large DNA fragments. A small fragment of DNA is most readily cloned into an 'insertion vector' that has a single target for a restriction enzyme in a non-essential part of is genome. Ideally, the genome of the vector will be small enough to allow insertion of DNA fragments of up to several kilobase pairs, and the target for the restriction enzyme will be in such a location as to give insertional inactivation of a gene and easy recognition of recombinants by their changed phenotype. DNA fragments of more than 10 kb can be cloned into a 'replacement vector', allowing the replacement of a non-essential fragment of vector DNA between two widely spaced targets for a particular restriction enzyme. The size-selection concept demands that the non-essential vector fragment be either retained or replaced, but not simply deleted. The presence of the dispensable fragment can often be detected by a simple phenotypic test or genetic selection, allowing recombinants to be clearly distinguished or selectively isolated from the parental vector. It is worth noting that each replacement vector places a lower as well as an upper limit on the size of DNA fragment that can be accommodated.

Maturation of recombinant phages

The replication of λ DNA switches fairly early in infection from the 'theta' mode, replicating monomeric circles, to the 'rolling circle' form that makes the multimeric concatenates that are the maturable DNA species. Intermediates in this switching process are sensitive to attack by a host cell nuclease, exonuclease V (Mackay & Linn 1974, 1976), the product of the *E. coli recB* and *recC* genes (Barbour & Clark 1970). Replicating λ DNA is protected against exonuclease V by the product of the phage's *gam* gene, a protein made early in infection that complexes with the nuclease and inhibits its action (Sakaki 3et al. 1973).

The *gam* gene of phage λ is located in the non-essential central region of the genome (Fig. D.1) and recombinants made with several replacement vectors will have lost this gene. Such recombinants will be unable to make concatameric DNA via rolling circles, the formation of which will remain susceptible to attack by exonuclease V. These phages can make maturable forms of DNA by recombination of monomeric molecules into multimeric circular forms, from which linear monomers can be excised and packaged. The required recombination events can be catalysed by the *E. coli* recombination enzymes or by the phage-coded generalized recombination system. Since the *red* genes of the phage that govern generalized recombination are adjacent to the *gam* gene (Fig. D.1), they will also be missing from recombinants made with some replacement vectors. Recombinants lacking both the *red* and *gam* genes will therefore be dependent on the *E. coli* recombination system for the generation of maturable forms of DNA. Lambda DNA is a poor substrate for the normal, *rec*A-dependent recombination pathway of *E. coli*, but can be considerably improved by acquisition of a *chi*-mutation (crossover host-spot instigator) (Lam *et al.* 1974) that generates a recombinogenic nucleotide sequence in the phage genome (Lam *et al.* 1974, Henderson & Weil 1975). Phages lacking the *red* and *gam* genes can also be propagated on *recB* or *recC* mutants of *E. coli* that do not produce exonuclease V.

Table D.1 Cloning capacities of the λ vectors

Phage	Enzyme	Insert size (kb)	Recognition of recombinants	Reference
λNM426	EcoRI	0–6.56	—	6
	EcoRI + SalI	0.8–13.8[a]	Spi[-]	
	EcoRI + XhoI	1.07–14.1[a]	Spi[-]	
λNM540	HindIII	0–11.4)8	—	6
	HindIII + XhoI	4.95–17.9[a]	Spi[-]	
λ NM641	EcoRI	0–11.6	Clear plaques	7
	XbaI	0–11.6[a]	—	
	EcoRI + XbaI	5.31–18.3[a]	Clear, Spi[-]	
λNM590	HindIII	0–11.3	Clear	7
λZEQS	EcoRI	0–8.5	Lac[-(b)]	9
λlac5-'Zi'	EcoRI	6.23–19.2	Lac[-], Spi[-(c)]	8
	SstI	0–8.5	Lac[-]	
λBV2	BamHI	0–12.6	—	4
λgt.λB'	EcoRI	2.13–15.1		11
Charon 4	EcoRI	7.02–20.0	Lac[-(c)]. Bio[-]	1
Charon 22	Xma	3.8–16.8	—	12
λNM781	EcoRI	2.16–15.2	Lac[-(d)]	7
λNM762	HindIII	2.57–15.6	Lac[-(d)]	7
λgtWES.75622	EcoRI	2.13–15.1	ColIb-insensitive[(e)]	2
λL47	EcoRI	8.37–21.4	Spi[-]	5
	HindIII	6.78–19.8	Spi[-]	
	BamHI	4.18–17.2	Spi[-]	
λ1059	BamHI	8.0–21.0	Spi[-]	3
λCh30	BamHI	6.1–19.1	Spi[-]	10
	HindIII	0–11.7	—	
	EcoRI	4.46–17.5	—	
	EcoRI + BamHI	7.24–20.2	Spi[-]	

Choice of vector

There is now a wide choice of phage vectors available and it is important to select the most suitable vector for the particular cloning exercise. The most important considerations in making that choice are probably the sequence-specificity of the endonuclease to be used and the size of the DNA fragments that are to be incorporated.

Insertion vectors allow the efficient cloning of small fragments of DNA, with an upper size limit of about 10 kb. Vectors are available for use with EcoRI (G$^\downarrow$AATTC), HindIII (A$^\downarrow$AGCTT) (Murray et al. 1977, Blattner et al. 1977, Pourcel & Tiollais 1977) and SstI (GAGCT$^\downarrow$C) (Pourcel & Tiollais 1977) that allow recognition of recombinant phages by their altered plaque-morphology or their Lac[-] phenotype on indicator plates. Insertion vectors are also available for use with BamHI (G$^\downarrow$GATCC) (Klein & Murray 1979, Rimm et al. 1980) or any of the enzymes that generate a 5' GATC-cohesive end. Lambda genomes carrying deletions can be used as vectors for fragments generated by enzymes that have a single site in a non-essential region of the phage genome, such as XbaI (T$^\downarrow$CTAGA) and XhoI (C$^\downarrow$TCGAG).

Replacement vectors impose both lower and upper limits on the size of the DNA fragment that can be cloned. In many cases the central, disposable fragment imports a readily recognized phenotype to the phage, allowing recombinants to be distinguished from the parental vector. In some cases the presence of the central fragment can prevent phage growth on certain host strains, so that recombinant phages can be directly selected. One such selection makes use of the fact that the growth of wild type λ is prevented on lysogens carrying the P2 prophage. (This phenotype is called Spi$^+$, for sensitive to P2 interference.) Derivatives of λ that lack both *red* and *gam* activity show the Spi$^-$ phenotype and grow on P2-lysogens (Zissler *et al.* 1971). Cloning into several λ vectors involves replacement of a central fragment containing the phage *red* and *gam* genes, allowing recombinants to be recognized or selected by growth on a P2-lysogenic host.

Choice of host strain

The most important feature of any potential host strain is that it allows the recombinant phages to grow well without imposing any undesirable selection of its own. Since the normal λ host, *E. coli K,* harbours a powerful restriction system, it is obviously important to use a non-restricting mutant.

Special host strains can be used to facilitate enrichment for recombinant phages. Thus a *pel*$^-$ host will inhibit growth of phages with small genomes (Scandella & Arber 1974), preferentially allowing propagation of phages with a full complement of DNA (Emmons *et al.* 1975). A strain of *E. coli* that shows an exaggeratedly high frequency of lysogenization by lambdoid phages (Lecocq & Gathoye 1973), probably mutant at the *hfl* locus (Belfort & Wulff 1971), has been used (Scherer *et al.* 1981) to select recombinants generated by insertion of DNA fragments into the *c*I gene of λ *imm*434 insertion vectors (Murray *et al.* 1977).

Lambdoid phages with small genome sizes (Bellett *et al.* 1971) or containing repetitive DNA sequences (Cooke & Hindley 1979) are known to be unstable in recombination-proficient hosts, where homologous recombination leads to duplication and deletion of repeated sequences. While the absence of *rec*A-dependent recombination will not guarantee stability of repetitive DNA (Brutlag *et al.* 1977, McClements *et al.* 1981) it will at least minimize potentially misleading events. The size-selection inherent in genome-packaging during phage propagation will also tend to stabilize those phages that have the optimal genome size.

Recombinant phages that have lost the *red* and *gam* genes

of phage λ cannot grow on *rec*A⁻ host strains, since their genomes are trapped in the non-maturable, monomeric circular form. Such phages can be grown on *rec*B or *rec*C mutants of *E. coli,* where the absence of the host's exonuclease V allows the replicating phage DNA to take on the rolling-circular state that is the normal substrate for DNA packaging (Enquist & Skalka 1973, Stahl *et al.* 1975). Although this stratagem minimizes recombination, it does make it difficult to obtain high-titre lysates.

References

Adhya, S., Gottesman, M. & De Crombrugghe, B. (1974) Release of polarity in *Escherichia coli* by gene N of phage λ: termination and autotermination of transcription. *Proc. natl. Acad. Sci. U.S.A.,* **71,** 2534.

Barbour, S.D. & Clark, A.J. (1970) Biochemical and genetic studies of recombination proficiency in *Escherichia coli,* I. enzymatic activity associated with $recB^+$ and $recC^+$ genes. *Proc. natl. Acad. Sci. U.S.A.,* **65,** 955.

Belfort, M. & Wulff, D.L. (1971) A mutant of *Escherichia coli* that is lysogenized with high frequency. In: *The Bacteriophage Lambda.* (Ed. Hershey, A.D.), pp. 739–742. Cold Spring Harbor Laboratories, Cold Spring Harbor, N.Y.

Bellet, A.J.D., Busse, H.G. & Baldwin, R.L. (1971) Tandem genetic duplications in a derivative of phage lambda. In: *The Bacteriophage Lambda.* (Ed. Hershey, A.D.), pp. 501–514. Cold Spring Harbor Laboratories, Cold Spring Harbor, N.Y.

Blattner, F.R., Williams, B.G., Blechl, A.E., Denniston-Thompson, K., Faber, H.E., Furlong, L., Grunwald, D.J., Kiefer, D.O., Moore, D.D., Schumann, J.W., Sheldon, E.L. & Smithies, O. (1977) Charon phages—safer derivatives of bacteriophage lambda for DNA cloning. *Science,* **194,** 161.

Brammar, W.J. (1982) Vectors based on bacteriophage lambda. In: *Genetic Engineering, Vol. 3.* (Ed. Williamson, R.), pp. 53–81. Academic Press, London.

Brookes, K. (1965) Studies on the physiological genetics of some suppressor–sensitive mutants of bacteriophage λ. *Virology,* **26,** 489.

Brutlag, D.L., Fry, K., Nelson, T. & Huey, P. (1977) Synthesis of hybrid bacterial plasmids containing highly repeated satellite DNA. *Cell,* **10,** 509.

Campbell, A. & Del Campillo-Campbell, A. (1963) Mutant of lambda bacteriophage producing a thermolabile endolycin. *J. Bacteriol.,* **85,** 1202.

Cooke, H.J. & Hindley, J. (1979) Cloning of human satellite III DNA: different components are on different chromosomes. *Nucl. Acid. Res.,* **6,** 3177.

Court, D. & Sato, K. (1969) Studies of novel transducing variants of λ: dispensability of genes N and Q. *Virology,* **39,** 348.

Couturier, M., Dambly, C. & Thomas, R. (1973) Control of development in temperate bacteriophages V. sequential activation of the viral functions. *Mol. Gen. Genet.,* **120,** 231.

Denniston-Thompson, K., Moore, D.D., Kruger, K.E., Furth, M.E. & Blattner, I.R. (1977) Physical structure of the replication origin of bacteriophage lambda. *Science,* **198,** 1051.

Dove, W. (1966) Action of the lambda chromosome. I. control of functions late in bacteriophage development. *J. Mol. Biol.,* **19,** 187.

Echols, H., Green, L., Oppenheim, A.B., Oppenheim, A. & Honigman, A. (1973) Role of the *cro* gene in bacteriophage λ development. *J. Mol. Biol.,* **80,** 203.

Emmons, S.W., MacCosham, V. & Baldwin, R.L. (1975) Tandem genetic duplications in phage lambda III. The frequency of duplication mutants in two derivations of phage lambda is independent of known recombination systems. *J. Mol. Biol.,* **91,** 133.

Enquist, L. & Skalka, A. (1973) Replication of bacteriophage λ DNA dependent on the function of host and viral genes I. interaction of red, *gam* and *rec*. *J. Mol. Biol.,* **75,** 185.

Franklin, N.C. (1974) Altered reading of genetic signals fused to the N operon of bacteriophage λ genetic evidence for modification of polymerase by the protein product of the N gene. *J. Mol. Biol.*, **89**, 33.

Harris, A.W., Mount, D.W., Fuerst, C.R. & Siminovitch, L. (1967) Mutations of bacteriophage λ affecting host cell lysis. *Virology*, **32**, 553.

Henderson, D.A. & Weil, J. (1975) Recombination-deficient deletions in bacteriophage λ and their interaction with *Chi* mutations. *Genetics,* **79**, 143.

Hendrix, R.W. (1971) Identification of proteins coded in phage lambda. In: *The Bacteriophage Lambda.* (Ed. Hershey, A.D.), pp. 355–370. Cold Spring Harbor Laboratory, Cold Spring Harbor, N.Y.

Joyner, A., Isaacs, L.M., Echols, H. & Sly, W. (1966) DNA replication and messenger RNA production after induction of wild-type λ bacteriophage and λ mutants. *J. Mol. Biol.*, **19**, 174.

Klein, B. & Murray, K. (1979) Phage lambda receptor chromosomes for DNA fragments made with restriction endonuclease I of *Bacillus amyloliquefaciens* H. *J. Mol. Biol.*, **133**, 289.

Lam, S.T., Stahl, M.M., McMilin, K.D. & Stahl, F.W. (1974) Rec-mediated recombinational hot spot activity in bacteriophage lambda II. a mutation which causes hot spot activity. *Genetics*, **77**, 425.

Lecocq, T.P. & Gathoye, A.M. (1973) *Arch. Int. Biochem. Physiol.*, **81**, 803.

McClements, W.L., Dhar, R., Blair, D., Enquist, L., Oskarsson, M. & Vande Woude, G.F. (1981) The long terminal repeats of integrated malony sarcoma provirus are like bacterial insertion sequence (IS) elements. Cold Spring Harbor Symp. Quant. Biol. XLV.

MacKay, V. & Linn, S. (1974) The mechanism of degradation of duplex deoxyribonucleic acid by the *rec*BC enzyme of *Escherichia coli* K12. *J. Biol. Chem.*, **249**, 4286.

MacKay, V. & Linn, S. (1976) Selective inhibition of the DNase activity of the *rec*BC enzyme by DNA binding protein from *Escherichia coli*. *J. Biol. Chem.*, **251**, 3716.

Murray, N.E. (1983) Phage lambda and molecular cloning. In: *The bacteriophage Lambda 2.* (Ed. Blattner, F.R.). Cold Spring Harbor Laboratory, Cold Spring Harbor, N.Y.

Murray, N.E., Brammar, W.J. & Murray, K. (1977) Lambdoid phages that simplify the recovery of *in vitro* recombinants. *Mol, Gen. Genet.*, **150**, 53.

Murray, N.E. & Murray, K. (1974) Manipulation of restriction targets in phage λ to form receptor chromosomes for DNA fragments. *Nature*, **251**, 476.

Pero, J. (1971) Deletion mapping of the site of action of the *Tof* gene product. In: *The Bacteriophage Lambda.* (Ed. Hershey, A.D.), pp. 599–608. Cold Spring Harbor Laboratory, Cold Spring Harbor, N.Y.

Pourcel, C. & Tiollais, P. (1977) λ p*lac*5 derivatives, potential vectors for DNA fragments cleaved by *Streptomyces stanfordii* restriction enzyme (S*stt*). *Gene*, **1**, 281.

Ptashne, M. (1971) Repressor and its action. In: *The Bacteriophage Lambda.* (Ed. Hershey, A.D.), pp. 221–237. Cold Spring Harbor Laboratory, Cold Spring Harbor, N.Y.

Radding, C.M. & Shreffler, D.C. (1966) Regulation of λ exonuclease II. Joint regulation of exonuclease and a new λ antigen. *J. Mol. Biol.*, **18**, 251.

Rambach, A. & Tiollais, P. (1974) Bacteriophage λ having *Eco*RI endonuclease sites only in the non essential region of the genome. *Proc. natl. Acad. Sci. U.S.A.*, **71**, 3927.

Rimm, D., Norness, D., Kucera, J. & Blattner, F.R. (1980) Construction of coliphage lambda charon vectors with *Bam*HI cloning sites. *Gene*, **12**, 301.

Sakaki, Y., Karu, A.E., Linn, S. & Echols, H. (1973) Purification and properties of the γ-protein specified by bacteriophage λ: an inhibitor of the host *Rec*BC recombination enzyme. *Proc. natl. Acad. Sci. U.S.A.*, **70**, 2215.

Scandella, D. & Arber, W. (1974) An *Escherichia coli* mutant which inhibits the injection of phage λ DNA. *Virology*, **58**, 504.

Schnös, M. & Inman, R.B. (1970) Position of branch points in replicating λ DNA. *J. Mol. Biol.*, **51**, 61.

Segawa, T. & Imamoto, F. (1974) Diversity of regulation of genetic transcription. II. specific relaxation of polarity in read-through transcription of the translocated *trp* operon in bacteriophage lambda *trp*. *J. Mol. Biol.*, **87**, 741.

Stahe, F.W., Crasemann, J.M. & Stahl, M.M. (1975) *Rec*-mediated recombinational hot spot activity in bacteriophage lambda. III. Chi mutations are site mutations stimulation *rec*-mediated recombination. *J. Mol. Biol.,* **94,** 203.

Thomas, M., Camberon, J.R. & Davis, R.W. (1974) Viable molecular hybrids of bacteriophage lambda and eukaryotic DNA. *Proc. natl. Acad. Sci. U.S.A.,* **71,** 4579.

Weil, J., Cunningham, R., Martin, R., Mitchell, E. & Bolling, B. (1972) Characteristics of λp3, a λ derivative containing 9% excess DNA. *Virology,* **50,** 373.

Williams, B.G. & Blattner, F.R. (1980) Bacteriophage lambda vectors for DNA cloning. In: *Genetic Engineering, Vol. 2.* (Eds Setlow, J.K. & Mullander, A.), pp. 201–281. Plenum Press, N.Y.

Zissler, J., Signer, E.R. & Schaefer, F. (1971) The role of recombination in growth of bacteriophage lambda I. The gamma gene. In: *The Bacteriophage Lambda.* (Ed. Hershey, A.D.), pp. 455–475. Cold Spring Harbor Laboratory, Cold Spring Harbor, N.Y.

Experiment D.1 Preparation of a bank of hybrid DNA

Background

The vector to be used in this exercise is a lambdoid phage, λL47, that can be used in conjunction with *Eco*RI (G$^\downarrow$AATTC), *Hin*dIII (A$^\downarrow$AGCTT) and *Bam*HI (G$^\downarrow$GATCC), or any enzyme that produces the same cohesive ends as any of the above. The tetranucleotide-recognizing enzyme *Sau*3A ($^\downarrow$GATC) is conveniently used in conjunction with *Bam*HI, since large fragments produced by partial digestion with *Sau*3A will have approximately random termini. Recombinants made using *Eco*RI, *Hin*dIII or *Bam*HI lose the central fragment of the vector that includes the *red* and *gam* genes, and therefore gain the ability to grow on a P2-lysogenic host strain. Recombinants can therefore be selected or scored on the basis of their so-called Spi$^-$ phenotype (sensitivity to P2-interference).

The efficiency of recovery of recombinants can be improved in several ways. Removal of the central fragment from the ligation reaction removes competition with the donor DNA. The preparation of donor DNA fragments* in the size-range of 15–20 kb has three effects: only recombinants with large inserts are recovered, small fragments that would compete with the large inserts are excluded, and fragments too large to generate viable genomes are eliminated.

(a) Preparation of the vector arms

Materials needed

λL47 DNA
10 × SalI buffer (1·5 M NaCl; 60 mM Tris-HCl, pH 7·9; 60 mM MgCl$_2$; 60 mM 2-mercaptoethanol)
SalI, BamHI
Phenol:chloroform:isoamylalcohol (25:24:1)
Chloroform:isoamylalcohol (24:1)
10 mM Tris-HCl, pH 7·5
2 M sodium acetate, pH 5·6
100% ethanol
70% ethanol
1·25 M NaCl; 10 mM Tris-HCl, pH 7·5; 1 mM EDTA
5 M NaCl; 10 mM Tris-HCl, pH 7·5; 1 mM EDTA

*See Appendix D.II and D.III for preparation of high molecular weight DNA.

TE buffer (10 mM Tris-HCl, pH 7·5; 1 mM EDTA)
Loading buffer (20% ficoll, 0·01% w/v bromo phenol blue)

Procedure

1 Restrict λ L47 DNA with 10×*Sal*I buffer (25 μl). DNA (50 μg), *Sal*I (50 u) and *Bam*HI (50 u). Add distilled water to make a final volume of 250 μl.

Incubate at 37°C for 2 h. Remove 1μl to analyse on a 0·8% agarose gel.

2 Terminate the reaction by mixing with an equal volume of phenol:chloroform:isoamylalcohol (25:24:1). Mix well, separate phases in a microfuge. Remove 250 μl from aqueous phase (top layer). Re-extract the phenol layer with 200 μl 10 mM Tris-HCl, pH 7·5. Mix well, separate phases in a microfuge. Remove 200 μl from aqueous phase (top layer). Remove phenol from the aqueous layers by extraction with an equal volume of chloroform:isoamyl alcohol (24:1). Mix well, separate phases in a microfuge and remove aqueous layers (approximately 450 μl). Mix well, with 50 μl 2 M sodium acetate, pH 5·6 and then mix with 1 ml ethanol. Precipitate the DNA in an ethanol/dry-ice bath for 5 min. Allow to thaw at room temperature, then pellet in a microfuge, for 5 min. Remove the ethanol layer carfully with a disposable micro pipette tip. Wash the DNA precipitates with 1 ml 70% ethanol, microfuge, 1 min. Dry the DNA pellet under vacuum 10 min. Resuspend the DNA in 500 μl 10 mM Tris-HCl, pH 7·5.

3 Prepare 17 ml NaCl gradients (1·25 M → 5 M NaCl in TE buffer) in ultracentrifuge tubes.

4 Load no more than 150 μg of digested DNA onto each gradient in a volume of 500 μl. Centrifuge the samples for 17 h, 20°C, 20 Krpm.

5 Collect 13 drop fractions (approximately 40–350 μl each)

6 Dilute fractions with an equal volume of TE buffer (to reduce NaCl concentration, otherwise large amounts of NaCl will appear in subsequent steps).

7 Add two vol of ethanol to precipitate DNAs. Microfuge each sample for 5 min. Wash DNA pellets with 1 ml 70% ethanol. Dry DNA pellets.

8 Resuspend each fraction in 20 μl TE buffer.

9 Prepare a 0·2% agarose gel (12 tracks).

10 Restrict 0·5 μg of λL47 DNA to be used as a marker on the following gel. Restrict with 10×*Sal*I buffer (1μl), DNA (0·5 μg), SalI (2 units) and *Bam*HI (2 units). Add distilled water to a final volume of 10 μl. Incubate at 37°C for 30 min. Add 2 μl 20% ficoll loading buffer, and load onto the 0·2% agarose gel, with the λL47 DNA fractions from the salt

gradient. Use 4 μl of sized DNA per track. Assay alternate fractions, i.e. 4, 6, ... 22.

11 Run gel at 30 V overnight.

(b) Preparation of genomic DNA—pilot partial digests

Materials needed

Genomic DNA
10×*Sau*3A buffer (500 mM NaCl; 60 mM Tris-HCl, pH 7·5; 50 mM $MgCl_2$)
Sau3A, *Hin*dIII, *Xho*I
0·2 M EDTA, pH 7·5
10×*Hin*dIII buffer (600 mM NaCl; 70 mM $MgCl_2$; 70 mM Tris-HCl, pH 7·4)
Termination buffer (20% ficoll, 20 μl; 0·2 M EDTA, pH 7·5, 5 μl; distilled water, 25 μl).

Procedure

1 Set up the following reaction mix: 10×*Sau*3A buffer (10 μl) genomic DNA (10 μg), distilled water to make a final volume of 100 μl. Pre-incubate at 37°C, 15 min.

Table D.2

	*Hin*dIII	*Xho*I
10×*Hin*dIII buffer	1 μl	1 μl
λ^+ DNA	0·5 μg	0·5 μg
Distilled water	to make up the volume to 10 μl	
Enzyme	1 unit	1 unit
Total	10 μl	10 μl

2 Prepare Eppendorf tubes to take samples. Into seven tubes place 5 μl of termination buffer.

3 Add 1 unit of *Sau*3A to the reaction mix and mix well. Continue to incubate at 37°C and remove 5 μl samples every 5 min (t = 0, 5, 10, 15, 20, 25, 30 min). Each sample should be placed immediately into a tube containing termination buffer.

4 Prepare a 0·2% agarose gel (12 tracks).

5 Prepare size markers for 0·2% gel. *Xho*I and *Hin*dIII digests of λ^+ DNA provide useful size markers. The *Xho*I digest generates three bands, 34 kb and 15 kb left and right-arms respectively, and a 49 kb band due to joining of the two arms at the *cos* ends. The *Hin*dIII digest generates nine bands, the largest three are most useful. These are 23 kb, 9·8 kb and left and right ends joined, 28 kb. Set up two reaction mixes as in Table D.2. Use 5 μl track.

6 Load 0·2% gel as in Table D.3. Run at 30 V overnight.

Table D.3

Track	Sample
1	λ*Xho*I
2	λ*Hin*dIII
3	0
4	5
5	10
6	15
7	20
8	25
9	30
10	λ*Xho*I
11	λ*Hin*dIII

(c) Preparation of genomic DNA—preparative partial *Sau*3A digests of genomic DNA

Procedure

1 Restrict genomic DNA in the following reaction mix: 10×*Sau*3A buffer (150 µl), genomic DNA (150 µg) and distilled water to make a final volume of 1·5 ml. Pre-incubate at 37°C, 15 min.

2 Add 15 units of *Sau*3A to reaction mix and mix well. Continue to incubate at 37°C, 20 min.

3 Terminate the reaction by mixing with an equal volume of phenol:chloroform:isoamylalcohol (25:24:1). Continue to recover the DNA in the same manner used for the vector DNA, λL47.

4 Resuspend the dry DNA pellet in 500 µl of 10 mM Tris-HCl, pH 7·5. Examine 2 µl on a 0·2% agarose gel with *Hin*dIII and *Xho*I digests of λ$^+$ DNA as size markers (use a track on the same gel used in the pilot experiment, part b).

5 Prepare a NaCl gradient in the same manner used to prepare λL47 vector arms. Never load more than 150 µg DNA per 17 ml polyallomer tube, otherwise resolution of DNAs of different sizes will be reduced. Centrifuge the NaCl gradients for 18 h, 18 kpm, 20°C, (alternatively, 3 h, 40 k, 20°C) using an AH627 rotor.

6 Set up fraction collector and collect 13-drop fractions (approximately 35).

7 Dilute each fraction with an equal volume of TE buffer (otherwise subsequent steps will release large quantities of NaCl, making recovery of DNA difficult).

8 Add 2 vol of ethanol. Leave in a ethanol/dry-ice bath, 5 min. Allow to thaw at room temperature. Pellet DNA in a microfuge, 5 min. Wash DNA pellet in 1 ml 70% ethanol, microfuge, 1 min. Dry DNA pellets. Dissolve DNAs in 40 µl 10 mM Tris-HCl, pH 7·5. This procedure de-salts DNA samples sufficiently to allow accurate sizes to be determined from a 0·2% agarose gel (high salt allows DNA molecules to migrate faster, so that the salt-gradients should distort the relationship between mobility and size).

9 Examine fractions (4, 6, ... 22) on a 0·2% agarose gel (12 tracks) with *Xho*I and *Hin*dIII digests of λ$^+$ DNA as size markers. Use approx 4 µl samples. Run at 30 V overnight.

(d) Ligation of donor and vector DNAs

Materials needed

1·0 M Tris-HCl, pH 7·5
0·2 M EDTA, pH 9·0
1·0 M MgCl$_2$
1·0 M 2-mercaptoethanol
0·1 M ATP, pH 7·0
T4 DNA ligase

Procedure

1   Prepare 10×T4 DNA ligase buffer:

1·0 M Tris-HCl, pH 7·5	66 μl
0·2 M EDTA, pH 9·0	5 μl
1·0 M MgCl$_2$	10 μl
1·0 M 2-mercaptoethanol	10 μl
0·1 M ATP, pH 7·0 (with NH$_4$OH)	1 μl

This is always prepared fresh.

2   Reaction:

λL47 arms	7 μg
Sized DNA	1·5 μg
10×T4 DNA ligase cocktail	5 μl
T4 DNA ligase	1 μl
H$_2$O to a final volume of 50 μl	

3   Incubate at 15°C, 4 h.
4   Place on ice to terminate the reaction. (The ligation mixture can be stored at −20°C.)

(e) *In vitro* packaging of ligated DNAs

Materials needed

Strain DB102 *metB supE supF hsdR$_K$ tonA trp^R*. For use as λ indicator pellet 10 ml of an overnight culture and resuspend in 10 ml of 10 mM MgSO$_4$.
BBL plates: BBL Trypticase 10 g; NaCl 5 gl^{-1}. When dissolved, aliquot out into 400 ml vol and mix each with 4 g of Difco Bacto Agar. Sterilize and pour
BBL top layer (sterile) per litre BBL trypticase 10 g; NaCl 5 g. When dissolved, aliquot out into 100 ml vol and mix each with 0·7 g of Difo Bacto Agar. Sterilize and store molten at 65°C. Before use, pre-warm at 48°C in a waterbath.
Thawed FTL and SE extracts on ice
Buffer A and B (see experiment D2 for details)
λ buffer: (6 mM Tris-HCl, pH 8·0; 10 mM MgCl$_2$; 100 mM NaCl; 0·5 mg ml^{-1} gelatine)

Procedure

1 Prepare buffers A and B, whilst packaging extracts are thawing on ice (approximately 1 h).

2 Reaction mix:

Buffer A	350 μl
DNA	50 μl
Buffer B	50 μl
Sonic extract (SE) see Experiment D2(a)	175 μl
Freeze-thaw lysate (FTL) see Experiment D2(b)	250 μl

3 Incubate for 90 min at room temperature.

4 Terminate the reaction with 11·5 ml of λ buffer. The large volume is necessary to dilute the packaging extracts which, at high concentration, inhibit the adsorption of packaged phage. To store add 50 μl $CHCl_3$ to packaged phage in a *glass* tube. Store at 4°C.

5 Spot-titre the packaged phage on a lawn of DB102. Mix 0·1 ml of DB102 indicator cells with 2·5 ml BBL top and pour the mixture onto a dry BBL plate. Allow agar to set, approximately 15 min. Meanwhile, make tenfold dilutions of packaged phage in λ-buffer. Use 1·5 ml Eppendorf tubes for dilution. Place 90 μl λ buffer into two tubes. Take 10 μl of packaged phage and mix with 90 μl λ buffer (1:10 dilution). Take 10 μl of this dilution and mix with a further 90 μl λ buffer (1:100 dilution). Spot 10 μl of each dilution (0, 1/10, 1/100 dilution) onto DB102 lawn. When spots have dried in, incubate at 37°C, O/N.

Experiment D.2 Preparation of extracts for *in vitro* packaging

Background

The *in vivo* packaging of phage genomes into λ heads can be mimicked *in vitro*, using concentrated crude cell-extracts made from induced λ lysogens. The packaging mixture is made by combining two cell extracts, each of which is deficient in one component of the phage maturation system. The major advantage of *in vitro* packaging is its relative efficiency: a good packaging reaction will yield greater than 10^8 phages/μg λ^+ DNA, which is at least 1,000-fold more efficient than transformation or transfection of *E. coli* cells.

In this experiment we will prepare crude cell-extracts for *in vitro* packaging (Scalenghe *et al.* 1981, Hohn 1979).

Materials needed

Strain BHB 2688: *rec*A sup° (λ*imm*434 *c*Its *b*2 *red*3 *Dam*15 *Sam*7)
Strain BHB 2690: *rec*A sup° (λ*imm*434 *c*Its *b*2 *red*3 *Eam*4 *Sam*7)
L-broth (sterile) (per litre: Bacto-tryptone, 10 g; yeast extract, 5 g; NaCl, 5 g; glucose, 1 g; adjusted to pH 7·0 with NaOH)
$CHCl_3$
Buffer A
   1 M Tris-HCl, pH 8·0   20 μl
   1 M $MgCl_2$ 3 μl
   β-mercaptoethanol   0·5 μl
   0·1 M EDTA, pH 7·0   10 μl
   Distilled water   966·5 μl
Buffer B
   1 M Tris-HCl, pH 7·5   6 μl
   pH 7·0, 0·1 M sperimidine—3 HCl   300 μl   neutralized
   pH 7·0, 0·2 M putrescine—2 HCl   300 μl   with Tris base
   1 M $MgCl_2$ 18 μl
   0·1 M ATP, pH 7·0, neutralized
   with NH_4OH 15 μl
   β-mercaptoetahanol   2 μl
   Distilled water   224 μl
   Liquid nitrogen
Tris-sucrose [10% sucrose (w/v); 50 mM Tris-HCl, pH 7·5]
Lysozyme (2 mg ml^{-1} in 0·25 M; Tris-HCl, pH 7·5)

References

Hohn, B. (1979) *In vitro* packaging of λ and cosmid DNA. In *Methods in Enzymology.* Vol. 68. (Ed. Wu, R.), pp. 299–309.

Scalenghe, F., Turco, E., Edström, J.E., Pirrotta, V. & Melli, M. (1981) Microdissection and cloning of DNA from a specific region of *Drosophila melanogaster* polytene chromosomes. *Chromosoma*, **32**, 205.

(a) Preparation of freeze-thaw lysate (FTL)

Procedure

1   Grow 3×250 ml cultures (5 ml of overnight culture of BHB2688 O/N per 250 ml L-broth) in L-broth at 30–32°C in 2 litre baffled flasks. At A_{630} = 0·6 (approximately 3 h after inoculation) remove 1 ml of cells for induction test. Transfer flasks to a 38–40°C incubator, add 250 ml of L-broth, pre-warmed to 62°C.

2   Grow at 38–40°C with vigorous shaking (200 rpm) for a further 1 h, to promote induction of phage. Finally, cool on ice and remove a further 1 ml sample.

3   Test for induction. Add a drop of chloroform to each sample of induced and non-induced cells. Cells induced will clear within 5 min.

4   Harvest cells in 6×250 ml rotor 9,000 rpm for 10 min at 4°C. Drain off all the supernatant, use sterile medical wipes to remove final traces.

5   Suspend pellets in 3 ml cold Tris-sucrose buffer, using 3 ml disposable plastic pipette. Split the suspension into two (approximately 3 ml each) and dispense into 2×10 ml polypropylene tubes. Add 75 μl of fresh lysozyme solution to each tube, mixing gently by inversion. Quick-freeze in liquid N_2. Thaw at room temperature, then place on ice (1 h) until completely thawed. Add 75 μl buffer B. Mix gently by inversion. Spin for 35 min, 33,000 rpm at 4°C, to clear. Remove the supernatant in 75 μl aliquots into pre-cooled screw-cap ampoules (approximately 30). Take care to avoid chromosomal DNA at bottom of tube. Quick-freeze this extract in liquid N_2. Store at −70°C. This extract is deficient in coat protein D and is sufficient for 14 reactions.

(b) Preparation of sonic extract (SE)

Procedure

1   Grow 250 ml culture of BHB2690 in L-broth at 30–32°C, in the same manner as BH2688. Also, induce in the same way.

2   Harvest in 6×250 ml rotor 9,000 rpm at 10 min, 4°C. Drain

off supernatant thoroughly, removing all liquid with medical wipes.

3   Suspend pellets in 1 ml buffer A, transfer to a single polypropylene tube and dilute with 2·6 ml buffer A. Take care not to create any foam as this will interfere with subsequent sonication.

4   Sonicate on ice without foaming (15 bursts of 5 s duration, with 30 s pauses) until no longer viscous. The length of sonication needs to be determined empirically. The shortest time usually gives the best results. At Leicester we have used an MSE sonicator at medium amplitude (7 μm).

5   Pellet cell debris at 6,000 rpm for 10 min at 4°C. With correct sonication the volume of the pellet should be minimal.

6   Place 55 μl aliquots into pre-cooled screw-cap ampoules and quickfreeze in liquid N_2. Store at $-70°C$. This extract is deficient in protein E, the major head component, and is sufficient for 14 reactions.

(c) Test packaging extracts

Procedure

1   Thaw FTL and SE on ice.
2   Mix successively in a 1·5 ml Eppendorf tube the following:

Buffer A	7 μl
λ^+ DNA	1 μl λ DNA provided (5 μg ml^{-1})
Buffer B	1 μl
SE	3·5 μl
FTL	5 μl

3   Microfuge for 3 s.
4   Incubate at room temperature for 90 min.
5   Dilute with 230 μl λ buffer.
6   The packaged lysate should have a titre of 2×10^6 to 1×10^7 phage ml^{-1}. Make serial dilutions of the phage preparation (10 μl phage+90 μl λ buffer) in λ buffer. Do this to obtain 1/10, 1/100 and 1/1,000 dilutions. Adsorb 10 μl of the 1/100 and 1/1,000 dilutions to 100 μl of DB102 plating cells, for 10 min at room temperature. Also, include a control, with cells but no phage. Add 2·5 ml molten BBL top-layer agar (pre-warmed at 48°C) and quickly pour onto dry BBL plates. Immediately swirl the plate to spread the top-layer evenly over the surface. Allow to set at room temperature (approximately 15 min) then incubate inverted at 37°C.
7   Score the plaques after overnight incubation. Determine efficiency of in vitro packaging (phage μg^{-1} λ^+ DNA) and percentage recovery of phage genomes.

Notes and trouble shooting

1 Efficiency with λ^+ DNA $2-5\times10^8$ phage μg^{-1} (1–2·5% recovery), at least 2–5,000-fold more efficient than transfection.

2 Temperature of incubator is critical. Initial growth must be at a temperature less than 32°C, otherwise growth will be slow. For induction the temperature must be near 38°C. At lower temperatures, the heat labile *cI*ts 434 repressor will be active and thus reduce the level of phage late gene expression. Temperatures greater than 40°C result in poor cell growth, consequently reducing the yield of late phage products.

3 Use sterile solutions, centrifuge pots, etc.

4 Take care to remove any trace of detergent from equipment. Its presence will result in premature lysis of cells during processing of extracts.

5 Prepare SE and FTL extracts on separate days for greatest efficiency.

6 Work quickly.

7 Use de-ionized, distilled water.

Appendix D.I Preparation of a gene-bank in a λ vector

Recombinant DNA methodology allows the fractionation of complex genomes into a series of DNA fragments that can readily be cloned into a suitable vector molecule. A sufficiently large population of recombinant clones to represent the complete sequence of the parental genome is known as a 'gene-bank'. In practice, gene-banks are most commonly constructed using either cosmid or phage vectors, largely because of the advantage of the relatively high efficiency of *in vitro* packaging compared with transformation of *E. coli*. The probability (P) that a given unique sequence is present in a collection of (n) clones is given by $P = 1-(1-L/M)^n$ (Clark and Carbon 1976), where L is the average size of fragments cloned and M is the genome size (all sizes in kilobases).

n can be computed using the following formula: $n = \log(1-P)/\log(1-L/M)$

For a mammalian genome of about 3×10^6 kb, assuming an average insert of 15 kb, a 99% probability ($P = 0.99$) of obtaining a given sequence requires approximately 10^6 recombinants.

The bacteriophage vector to be used here is λL47 (Loenen and Brammar 1980). The vector DNA cut with BamHI will serve as receptor for DNA fragments generated by the tetranucleotide-recognizing enzyme Sau3A, which generates the same cohesive termini as BamHI. The ability to clone large DNA fragments produced by partial digestion with Sau3A having a tetranucleotide recognition site and therefore making frequent breaks, effectively generates a random collection of genomic DNA fragments. The appropriate number of recombinants should then generate a gene-bank containing most, if not all, sequences from the organism of choice.

References

Clark, L. & Carbon, J. (1975) Biochemical construction and selection of hybrid plasmids containing specific segments of the *Escherichia coli* genome. *Proc. natl. Acad. Sci. U.S.A.*, **72**, 4361.

Loenen, W.A.M. & Brammar, W.J. (1980) A bacteriophage lambda vector for cloning large DNA fragments made with several restriction enzymes. *Gene*, **10**, 249.

Appendix D.II DNA extraction from mouse livers

Materials required

Solutions (all ice-cold)

SE buffer (150 mM NaCl; 100 mM EDTA, pH 8·0)
Lysis mix
  2% SDS, 8% Tris-isopropyl naphthalene sulphanate
  Dissolved in 12% butanol
5 M NaClO$_4$
Phenol/chloroform/isoamylalcohol (25:24:1)
Phenol distilled and equilibrated against distilled water
10 mM Tris-HCl, pH 7·5
TE buffer (10 mM Tris-HCl, pH 7·5; 1 mM EDTA)
10×TNE buffer (500 mM Tris-HCl, pH 7·5; 1 M NaCl; 50 mM EDTA)

Procedure

1 To 1 g of liver add 5 ml of SE and homogenize gently (approximately 15–30 s) with a variable speed motor-driven-teflon-on-glass stirrer.
2 Homogenize.
3 Tip into a 100 ml plastic beaker containing 5 ml of lysis mix, on ice. Mix gently by swirling (approximately 2–3 min).
4 Add 2·5 ml of 5 M NaClO$_4$. Swirl on ice for approximately 10 min until an emulsion forms.
5 Pour into a 100 ml plastic beaker containing 6·25 ml of phenol. Swirl gently on ice until an emulsion forms (approximately 15 min).
6 Pour into a 30 ml siliconized cortex tube and spin HB4, 10,000 rpm for at 20 min, 4°C.
7 Remove upper layer into a 250 ml plastic beaker (on ice) using a cut-off micropipette tip.
8 Re-extract phenol layer (bottom) with 2·5 ml of 10 mM Tris-HCl, pH 7·5 (steps 5–7).
9 Pool aqueous layers (approximately 15 ml).
10 Add 30 ml of ethanol.
11 Swirl until DNA clumps. Decant off ethanol layer. Rinse with 50 ml of 70% ethanol. Decant off ethanol.
12 Resuspend DNA in 0·1×TNE 1·5 ml in a 15 ml siliconized Corex tube.

13 Add 20 µl of RNase (20 mg ml^{-1}, pre-heated at 90°C for 10 min to destroy residual DNase activity). Incubate at 37°C for 15 min.
14 Add 150 µl 10% SDS.
15 Add 165 µl 10×TNE.
16 Add 40 µl 20 mg ml^{-1} pronase (pre-digested for 1 h at 37°C before use).
17 Incubate at 37°C for 15 min.
18 Make volume up to 3 ml with 10 mM Tris-HCl, pH 7·5 (approximately 1·25 ml).
19 Add 3 ml of phenol. Mix gently in a 100 ml plastic beaker. Tip into 30 ml siliconized Corex tube. Centrifuge at 10,000 rpm for 10 min at 4°C. Remove upper layer in a 100 ml plastic beaker, on ice.
20 Re-extract phenol layer with 1 ml of 10 mM Tris-HCl, pH 7·5. Repeat step 19.
21 Pool aqueous layers.
22 Add 0·4 ml 2 M NaAc, pH 5·6. Mix gently (to ensure formation of Na-DNA salt, after ethanol precipitation).
23 Add 9 ml of 100% ethanol. Swirl gently (on ice) until DNA clumps. Decant off ethanol.
24 Resuspend DNA in 4 ml of 10 mM Tris-HCL, pH 7·5. DNA should dissolve within minutes provided precipitate has not been compacted by handling.
25 Add 0·4 ml 2 M NaAc, pH 5·6. Mix gently.
26 Add 9 ml of 100% ethanol. Swirl gently (on ice) until DNA clumps. Decant off ethanol.
27 Rinse DNA with 70% ethanol (10 ml). Decant off ethanol.
28 Dry DNA. Resuspend in 1 ml of 10 mM Tris-HCl, pH 7·5.
29 DNA 500–1,000 µg ml^{-1}. Size greater than 100 kb.
30 *May* require dialysis against 10 mM Tris-HCl, pH 7·5 if too much salt is present.
31 Dissolve the DNA in a small volume of 0·1×TNE. Add an equal volume of 2·5 M phosphate (pH 8·0) and mix. Add the same vol of 2-methoxyethanol and mix. Centrifuge at 10,000 rpm for 2 min.
32 Carefully remove the viscous upper phase containing DNA.
33 Re-extract the lower oily phase by adding 1 vol H$_2$O and 1 vol K phosphate and mixing. Add 1 vol 2-methoxyethanol, mix and centrifuge (this extraction removes carbohydrate impurities and is very useful for obtaining good-quality clean DNA).
34 Dialyse the pooled upper phases against 10 mM Tris-HCl (pH 7·5) at 4°C overnight and store the dialysed DNA at −20°C.

References

Jeffreys, A.J. & Flavell, R.A. (1977) A physical map of the DNA regions flanking the rabbit β-globin gene. *Cell*, **12**, 429.

Jeffreys, A.J. & Flavell, R.A. (1977) The rabbit β-globin gene contains a large insert in the coding sequence. *Cell*, 12, 1097.

Appendix D.III Isolation of high molecular weight bacterial DNA

Procedure

1 Wash 200 ml stationary cells twice with an equal volume of ice-cold 0·01 M Tris-HCl, pH 7·9; 0·001 M EDTA; 0·1 M NaCl.
2 Resuspend in 25 ml same buffer.
3 Add 5 ml, 10 mg ml^{-1} lysozyme and incubate at 37°C for 10 min.
4 Add 30 ml buffer containing 2% sarcosyl NL97 and 20 μg ml^{-1} RNAase A.
5 Incubate at 42°C for 60 min.
6 Add pronase to a final concentration of 1 mg ml^{-1} and incubate at 42°C for 4 h (pronase self-digested at 37°C for 1 h before use).
7 Extract lysate three times for 30 min with phenol (equilibrated against 0·5 M Tris-HCl, pH 8·0).
8 Precipitate aqueous phase with ethanol.
9 Resuspend in 20 ml 0·01 M Tris-HCl, pH 7·4; 0·001 M EDTA. (High molecular weight DNA takes a long time to dissolve, 3 days, 4°C.)
10 Dialyse against same buffer, with three changes.
11 DNA *ca* 200 kb, *ca* 4 mg 200 ml^{-1}.

References

Chow, L.T., Kahmann, R. & Kamp, D. (1977) Electron microscopic characterisation of DNAs of non-defective deletion mutants of bacteriophage Mu. *J. Mol. Biol.*, **113**, 591.

Section E Electronmicroscopy of nucleic acids

Pam McTurk

Introduction

The visualization of nucleic acid molecules in the electron microscope (EM), originally developed by Kleinschmidt, has become a powerful approach to the analysis of nucleic acid structure, being particularly suitable for the study of cloned segments of DNA. Accurate, quantitative measurements can routinely be achieved with very small quantities of nucleic acid. The concentration, size, and physical state of a preparation can be assessed, and information obtained on the degree and type of contamination. Regions of homology and non-homology between related molecules can be measured and the sizes and positions of insertions, deletions, inversions and translocations can be revealed by heteroduplex mapping. Hybrids between DNA and RNA can also be visualized, giving a valuable method (R-loop mapping) for determining the position of a coding sequence within a region of DNA and the number, sizes and positions of introns and exons in a cloned structural gene.

The exercises described in this section should serve as an introduction to the basic techniques of electron microscopy of nucleic acids, enabling one to carry out routine analyses of DNA and RNA molecules, heteroduplex mapping and R-loop mapping. Procedures for studying the interaction of proteins with DNA molecules can be obtained from the references.

In Section E.1 some general procedures including general laboratory equipment, preparation of various types of grid, general spreading and staining techniques, use of the EM and associated photography and finally the sizing of nucleic acids will be covered.

References

Bakken, A.H. & Hamkalo, B.A. (1978) Techniques for visualisation of genetic material. In: *Principles and Techniques of Electron Microscopy. Vol. 9.* (Biological appl) (Ed. Hayat, M.A.). van Nostrand-Reinhold, New York.

Davis, R.W., Simon, M. & Davidson, N. (1971) Electron microscope heteroduplex methods for mapping regions of base sequence homology. *Meth. Enzym.* **21**, 413.

European Molecular Biology Organisation (1979) *Laboratory Course on Recombinant DNA.* Supplement: Electron Microscopic Analysis of DNA molecules. Organiser, Bernadi, G.

Ferguson, J. & Davis, R.W. (1978) Quantitative electron microscopy of nucleic acids. In: *Advanced Techniques in Biological Electron Microscopy. Vol. 11.* (Ed. Kochler), pp. 123–171. Springer-Verlag, Heidelberg, New York.

Fischer, H.W. & Williams, R.C. (1979) Electron microscopic visualization of nucleic acids and of their complexes with proteins. *Ann. Rev. Biochem.*, **48,** 649.

Kleinschmidt, A.K. (1968) Monolayer techniques in electron microscopy of nucleic acid molecules. *Meth. Enzym.*, **12,** 361.

Lang, D. & Mitani, M. (1970) Simplified quantitative electron microscopy of biopolymers. *Biopolymers,* **9,** 373.

Thomas, J.O. (1978) Electron microscopy of DNA. In: *Principles and Techniques of Electron Microscopy. Vol. 9.* (Ed. Hayat, M.A.). van Nostrand-Reinhold, New York.

General EM procedures

(a) Laboratory equipment and materials

1 Clean glassware and suitably filtered solutions are essential. Initially, the glassware should be acid-washed for several hours in 50% (v/v) nitric acid: 50% sulphuric acid, then very thoroughly rinsed. After use, rinse it in distilled water and keep it separate from general laboratory glassware.

2 Optical glass microscope slides, used to spread nucleic acids, are stored in distilled water in a glass-staining trough (Gallenkamp HJ L-480W). Before use, these are acid-washed and thoroughly rinsed.

3 Spreading DNA is difficult on a draughty bench and either a quiet, sheltered bench area or a draught-free hood should be used.

4 Best quality glass-distilled water should be used.

5 Number 5 fine crossover electron microscope tweezers are recommended for handling grids (Agar Aids T539).

(b) Preparation of electron microscope grids

300 mesh copper grids are usually used. These are either handled (Emscope Laboratories Ltd, Veco E.M. grids V320) or plain (Agar Aids G2300), depending on the preferred method for mounting the DNA and the type of vacuum coating unit available. The grids should be washed in acetone and air-dried before coating. Grids are best stored on filter paper in glass petri dishes (98 mm diameter).

(i) Collodion-coated grids

For all routine work, collodion-coated grids can be used and these are freshly prepared each week. The film is made from 0·25% collodion in amyl acetate (Polaron 2290/3), which must be kept free from water. Clean a glass slide with alcohol. Then, using a Pasteur pipette, place a few drops of collodion on the slide and allow it to form an even film. Rest the slide against a beaker whilst the surplus drains away and the film dries. Score the slide round the edge with a scalpel. Fill a crystallizing dish with distilled water and gently float the film off the slide onto the surface of the water. Place the grids on the film (matt side down) and pick both up with a strip of Nescofilm. Place this on filter paper in a petri dish and dry under a bench light. Store the coated grids in a vacuum desiccator.

Carbon can be directly deposited on the reverse side of collodion grids to increase stability in the beam. The experimental detail will depend on the vacuum-coating until being used. Place a narrow strip of double-sided tape on a glass slide and put the collodion coated grids, coat side down, onto the slide and just touching the tape. Put the slide in the coating unit with a piece of filter paper stuck beside it and the carbon jig about 10 cm above it. Evaporate the carbon with a good vacuum to ensure a fine grain and stop when the filter paper is pale grey.

(ii) Carbon-coated grids

The indirect deposition of carbon produces a stronger and finer film, but it is less hydrophilic. The carbon is deposited on mica and then floated off onto the grids.

Cleave a thin sheet of mica 25×25 mm (Agar Aids G250-2) using EM forceps. Place this, cleaved side up, on a glass petri dish lid, the edges touching double-sided tape. Put in the coating unit 10 cm beneath the carbon jig. Evaporate the carbon as before but until the filter paper is mid-grey.

Put a wire mesh tray with handles in a crystallizing dish filled with distilled water. Place the grids, matt side up, on the tray. Slowly float the carbon off the mica using a shallow angle. If this is difficult, cut one edge off the mica. Draw the grids up beneath the carbon film and allow to air dry.

To make carbon grids hydrophilic, rinse for 10–15 seconds in 95% ethanol just before use. Then rinse twice in the solution used to spread the nucleic acid. Pick up the nucleic acid at once. Care must be taken not to destroy the carbon film (personal communication, Steve Hill).

(iii) Holey carbon film

A holey carbon grid is needed to check astigmatism on the microscope. Prepare a collodion-coated slide and, whilst still wet, huff strongly on it. Make the collodion-coated grids as previously described and then directly deposit carbon onto the coated side of the grid.

Reference

Willison, J.H.M. & Rowe, A. (1980) *Practical Methods in Electron Microscopy. Vol. 8.* (Ed. Glauert, A.M.). North-Holland Publishing Co., Amsterdam.

(c) Spreading nucleic acids for mounting on grids

The experiments require that DNA be spread as a monomolecular layer on the surface of a solution. In most experiments the DNA is present in the hyperphase, with cytochrome C as the spreading agent. This hyperphase is then allowed to run over the surface of a suitable hypophase and the DNA picked up with EM grids.

(i) Cytochrome C as a spreading agent

Nucleic acid molecules are very thin and tend to have an unstable, elastic structure in solution. Cytochrome C stabilizes the molecules so that they can be measured, and increases their diameter making them easier to visualize. Cytochrome C forms a basic surface monolayer which absorbs the nucleic acid. It is used in nucleic acid spreading techniques and in the aqueous droplet method. The amount of cytochrome C bound to DNA depends on the salt concentration. Too much cytochrome C will obscure small loops but too little prevents the formation of a monomolecular film.

(ii) Formamide

Formamide is used when it is necessary to visualize single-stranded DNA, which in aqueous solution appears clumped. Formamide affects the denaturation of DNA and combined with the correct temperature and pH enables the visualization of deletion and substitution loops in heteroduplex experiments, denaturation bubbles and R-loops.

Materials needed

Acid-washed glass slides stored in water. Plastic petri dish lid (100 mm^2) or other shallow, hydrophobic container. Talcum powder and fine paint brush. Gilson micropipette to dispense hyperphase. EM forceps. Coated grids. Filter paper. Stopwatch.

Procedure

1 Place a clean, drained slide at a fairly shallow angle (about 20° to horizontal) in the petri dish.
2 Pour the hypophase down the glass slide and into the dish and allow it to settle and the slide to drain.

3 Sprinkle talcum powder, with a brush, across the hypophase surface, half-way between the slide and the opposite edge of the dish.

4 Slowly pipette the hyperphase containing the DNA, allowing it to run down the glass slide and onto the surface of the hypophase. Hold the pipette about 1 mm above the surface of the slide and 1 cm from the edge of the hypophase. The edge of the hyperphase pushes the talcum powder back and the surface film can be seen as it travels over the hypophase.

5 In heteroduplex mapping, leave for 1 min for small molecules and 2–3 min for larger ones before picking up the DNA. When R-loop mapping, pick up the molecules at once.

6 Dry the tips of the forceps before picking up the DNA. With the coated surface of the grid face down, pick it up by the edge (or the tab if used) and bend slightly to enable the grid to touch the hypophase surface horizontally. This will pick up a droplet containing DNA. If sufficient DNA is present, the grid will hold a droplet with a sizeable miniscus. Generally, pick up DNA from an area 1 cm away from the glass slide.

7 Drain the edge of the grid on filter paper.

8 Stain and rinse.

9 Repeat this until sufficient grids are prepared.

10 Air-dry and shadow.

(d) Staining and shadowing nucleic acids for electron microscopy

The DNA should show sufficient contrast to be visualized, but not be so coarsely stained and overshadowed as to make interpretation difficult. For routine work grids are stained with uranyl acetate and then rotary shadowed with metal. Phosphotungstic acid will produce excellent contrast but is extremely coarse and only of limited use.

(i) Uranyl acetate stain

The stock solution of stain is made up monthly and stored in the cold and dark.

Materials needed

Uranyl acetate (BDH 10288)
1 M HCl
Distilled H_2O
90% ethanol, millipore (EW HP) filtered

Procedure

1   To make a stock solution of uranyl acetate, take 0·212 g uranyl acetate, add 5 ml 1 M HCl and make up to 100 ml with distilled H_2O. Millipore (HAWP) filter.
2   To dilute this for use, take 9·5 ml 90% ethanol and add 0·5 ml stock solution of stain.
3   Stain the grids by gently waving in the solution for 20–30 s.
4   Rinse in 95% alcohol (EWHP filtered) for 2–3 s.
5   Air dry grids on filter paper.

(ii) Rotary shadowing

In order to visualize nucleic acids easily they are shadowed with an evaporated metal. This step does not always give consistent results. The method used largely depends on the equipment available. The grids are rotary-shadowed at a low angle of 5–7°. The evaporation of 1·5 cm of 0·2 mm Pt/Pd wire (Agar Aids E412) wound round an electrode made of 1 mm tungsten wire (Agar Aids E407-1) and placed 9 cm from the grids is found satisfactory. It is more difficult to evaporate platinum, but this gives a fine grain and high contrast.

C/Pt shadowing gives excellent results but is not always practical. In this laboratory we now use a Polaron microsputter module with the low angle shadowing attachment. The ion beam gun has a gold target and the specimens rotate at an angle of 5°C.

References

Kay, D. (1965) *Techniques for Electron Microscopy*, 2nd edn., Blackwell Scientific Publications, Oxford, London, Edinburgh.
Willison, J.H.M. & Rowe, A. (1980) *Practical Methods in Electron Microscopy. Vol. 8.* (Ed. Glauert, A.M.). North Holland Publishing Co., Amsterdam

(e) Use of the electron microscope

To maximize the quality and stability of the image, the microscope must be used carefully and the alignment checked daily. Special attention should be given to astigmatism and this corrected with a holey carbon film. Aim to achieve a high contrast image which is not beam-damaged.

Reducing the accelerating voltage will increase contrast but also specimen damage. A recommended compromise is 60 kV. A standard condenser aperture of 200 μm and small objective aperture of 40 μm are preferred. Always use liquid nitrogen to reduce specimen damage. If a side-entry specimen chamber with goniometer is used, the eucentric position must be corrected for each grid. Search the grid, with the aid of

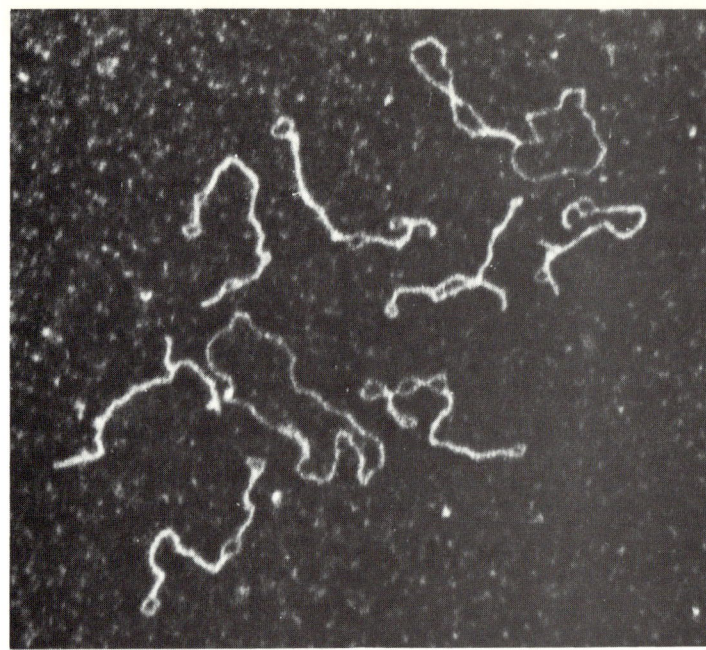

Fig. E.1 DNA of the plasmid pBR322 (4,362 base pairs) showing supercoiled and open circle forms. Aqueous droplet technique. Platinum, palladium rotary shadowed. Dark field microscopy, Jeol 100 cx, A.V., 80 kV.

binoculars, using a magnification on the microscope between 3,000 and 6,000.

Dark field microscopy (Fig. E.1) produces excellent results but can be difficult to set up and may cause excessive beam damage. Scanning transmission electron microscopy (STEM)

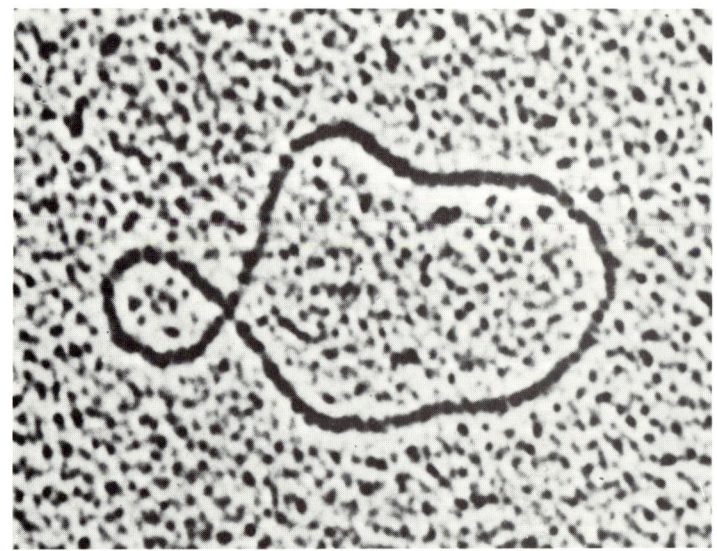

Fig. E.2 Open circle of the plasmid pBR322 DNA (4,362 base pairs). Polaron low angle, ion beam, gold shadowed. Jeol 100 cx, STEM mode. A.V. 100 kV. (×60,000).

can be used to enhance contrast artificially, but the resolution is reduced and searching the material is difficult (Fig. E.2).

Reference

Dubochet, J., Ducommun, M., Zollinger, M. & Kellenberger, E. (1971) A new preparation method for dark field electron microscopy of biomacromolecules. *J. Ultrastruct. Res.*, **35**, 147.

(f) Photography

A good preparation, suitable for analytical work should have nucleic acids at a suitable concentration to avoid overlapping molecules and yet be sufficiently frequent to reduce tedious searching. Well spread DNA lies in gentle curves, single-stranded DNA being thinner and having a less extended appearance. DNA molecules which are all lying in one direction or have an overstretched appearance are unsuitable, as are crinkled or clumped molecules. In one experiment, all molecules photographed for measurement should be taken from the same grid and grid bars must be avoided.

For each experiment all the electron micrographs of nucleic acid molecules, including marker molecules, should be taken from the same grid. Good quality film should be used or blotching may occur. The electron micrographs are all taken at the same magnification, using the lowest magnification compatible with seeing the molecule on the negative (phage λ at 8 k, pBR322 at 10–20 k). With the enlarger set in one position the negatives should be printed on hard paper. To increase the contrast of the print a long exposure time is used.

(g) Length measurement of nucleic acid molecules

The experiments to be outlined should give length measurements accurate to within 100–200 base pairs. At least 20 molecules are needed to give an accurate measurement of size.

If internal standards are not present within the molecules to be measured, DNA (marker) molecules of known size will need to be added and a similar number measured. If necessary, add both single-stranded (ϕx174) and double-stranded (pBR322, open circle) molecules. Avoid markers that are a similar size to the molecules in the experiment. Approximately 0·05 µg of marker molecules should be added to the hyperphase just before spreading, maintaining the overall concentrations of reaction mixture DNA, formamide and buffer. In the aqueous droplet technique marker molecules can be added when the DNA solution is prepared. DNA markers should be in open circle form.

To measure nucleic acid molecules a planimeter (MOP) is used, the prints being measured directly with a computerized pen.

Experiment E.1 Mounting double-stranded DNA using the aqueous droplet technique

This is used to check quality, condition (supercoiled, open circle or linear) and concentration of double-stranded DNA. It is also a useful check before a heteroduplex experiment. This droplet technique is quicker to use than spreading techniques.

Materials provided

Ammonium acetate (2 M NH_4 acetate, 0·01 M EDTA, pH 7). Millipore filterd and store at 4°C. Used at room temperature before use.

Cytochrome C. Type VI from horse heart (Sigma 07752). Stock solution of 0·4 mg ml^{-1}. Filtered and stored in freezer in small amounts. Use at room temperature and discard remainder. Dilute 1 in 4 in H_2O before use (0·01% w/v final conc.).
Uranyl acetate stain
Plastic petri dish

Procedure

1 Mix 2·56 ml H_2O, 0·3 ml 2 M ammonium acetate, 0·01 M EDTA, pH 7·0 and 0·08 ml 0·01% cytochrome C.
2 Mix 0·25–1 ml of the above solution with the DNA. For 0·25 ml of solution add: 50 μl DNA at 1 μg ml^{-1}, 2–5 μl DNA at 10 μg ml^{-1} and 0·5 μl DNA at 100 μg ml^{-1}.
3 Place several small droplets (approximately 50 μl) of the mix on a plastic petri dish. Leave for 30–45 min for the DNA/cytochrome C to concentrate on the surface.
4 Using a coated grid, pick up DNA from the surface of the droplet (1–2 grids per droplet).
5 Stain and rinse grids.
6 Store in a vacuum desiccator until shadowed.

Experiment E.2 Heteroduplex mapping

In heteroduplex mapping regions of non-homology between different DNA molecules can be seen (Figs E.3 and E.4). Initially, the two types of double-stranded DNA are denatured and then renatured in the presence of formamide. Treatment for 30 min at 25°C has been found satisfactory for most molecules.

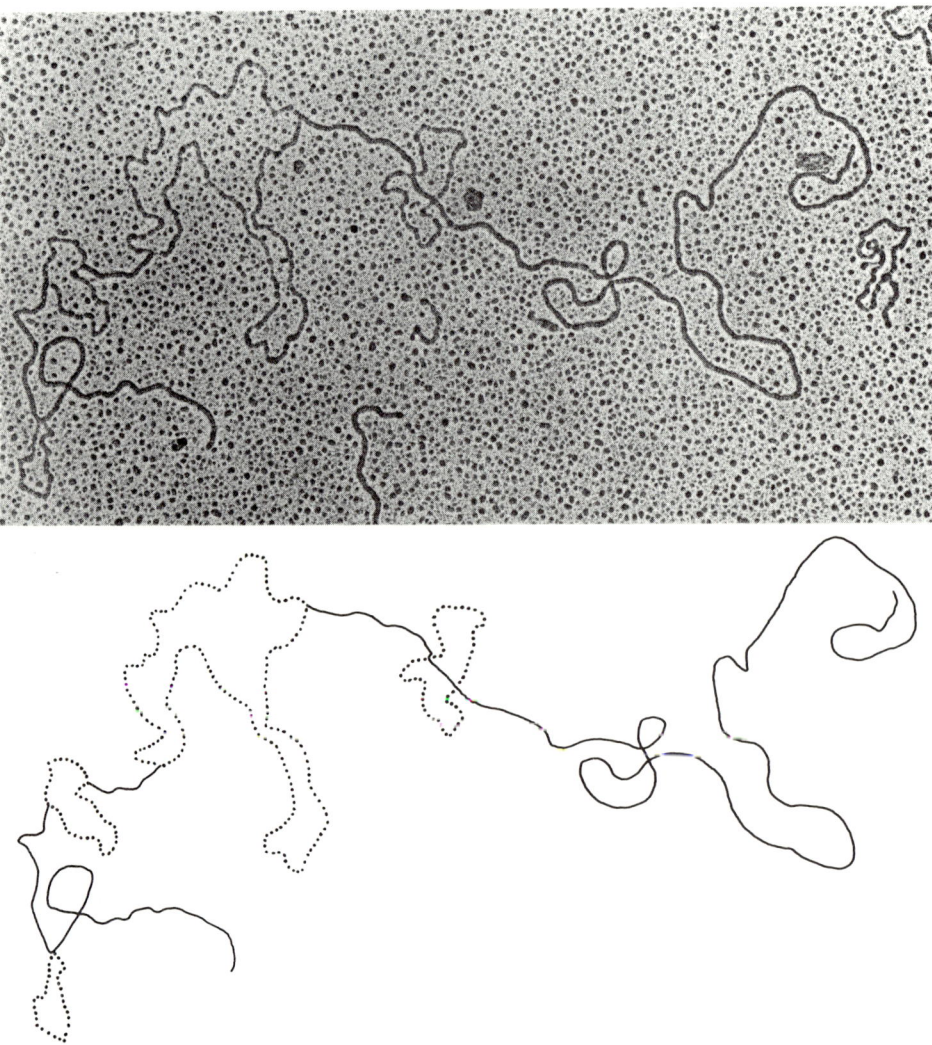

Fig. E.3 Heteroduplex between the DNAs of wild-type bacteriophage λ (48,500 base pairs) and a derivative, showing two single-stranded deletion loops and two single-stranded substitution loops. Polaron low angle, ion beam, gold shadowed. Jeol 100 cx, A.V., 60 kV.

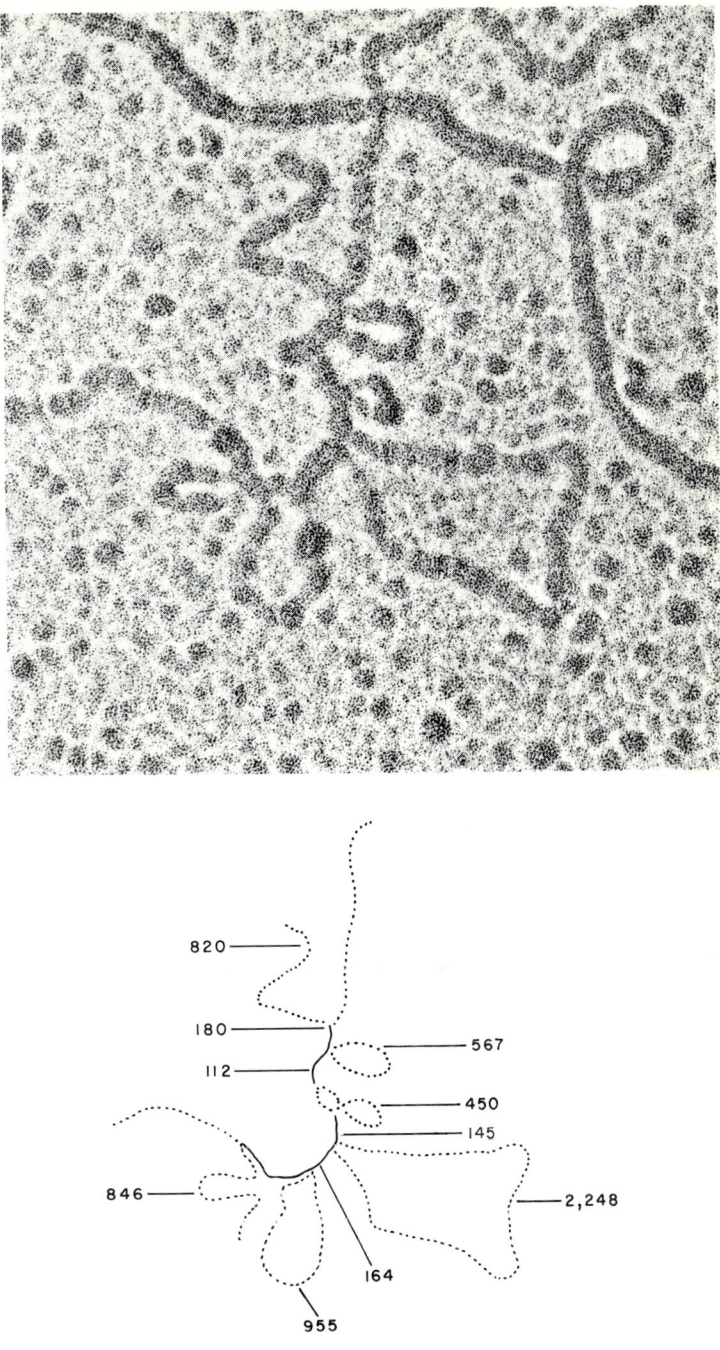

Fig. E.4 Heteroduplex between a cDNA copy of mouse renin mRNA and a cloned renin gene, showing introns and exons, with approximate sizes in base pairs. Polaron low angle, ion beam, gold shadowed. Jeol 100 cx, A.V., 60 kV.

Materials needed

Formamide specially purified (BDH 44 254 2Y). If problems occur, deionize the formamide with Amberlite monobed resin MB-1 (BDH 55007)

Hyperphase buffer (1 M Tris, 0·1 M EDTA, pH 8·5)
Reaction mix buffer (0·1 M Tris, 0·01 MEDTA, pH 8·5)

Cytochrome C. Stock solution of 0·4 mg ml^{-1}.
Uranyl acetate stain.
75°C waterbath
25°C waterbath

Procedure

1 In an Eppendorf tube mix 28 μl formamide, 2·6 μl NaCl, 4 μl of reaction mix buffer and the DNAs at a combined concentration of 1–3 μg ml^{-1} in a total volume of 5·4 μl.

2 Place the Eppendorf tube in a water bath for five min at 75°C and then 30 min at 25°C.

3 To prepare the hyperphase, remove the reaction mix from the water bath and dilute, if necessary, using a solution of the reaction mix in which the DNA has been replaced by water. 20 μl of solution should contain a total of 0·03–0·01 μg DNA for large (40 kb) molecules, 0·05 μg DNA for smaller (10 kb) molecules and can be used undiluted for small (5 kb) molecules.

4 To spread the hyperphase, take the correct concentration of reaction mix and DNA size markers in 29 μl and add 36 μl formamide, 10 μl hyperphase buffer and finally 25 μl 0·4 mg ml^{-1} cytochrome C. Spread on a deionized water hypophase.

Reference

Durnham, D.M., Perrih, F., Gannon, F. & Palmiter, R.D. (1980) Isolation and characterization of the mouse metallothionein—I gene. *Proc. natl. Acad. Sci. U.S.A.*, **77**, 6511.

Experiment E.3 Mapping self-annealing molecules (snap-backs)

This method is used to find inverted repeated sequences of the type occurring in transposons or IS elements. Duplex DNA molecules are denatured and single-stranded molecules will rapidly re-anneal in places of inverted homology. The snap-backs show as double-stranded regions associated with a single-stranded loop. Both double-stranded and single-stranded markers should be used but it can be less confusing if two experiments are run, one with a double-stranded marker and the other with a single-stranded one.

Materials needed

The heteroduplex protocol is followed and the initial mix should contain not more than 0·3 µg DNA in 20 µl.

Procedure

1 Denature solution containing DNA for 5 min at 75°C.
2 Reanneal for 10–20 min at 25°C.
The time taken to reanneal depends on the distance apart of the inverted repeat sequences.
3 Molecules are spread and mounted and can be picked up at once.

References

Coggins, L.W., Grindley, G.J., Vass, J.K., Slater, A.A., Montague, P., Stinson, M.A. & Paul, J. (1980) Repetitive DNA sequences near 3 human β type globin genes. *Nucl. Acid. Res.,* **8,** 3319.

Ohtsubo, H., Nyman, K., Doroszkoewicz, W. & Ohtsubo, E. (1981) Multiple copies of iso-insertion sequences of ISI in *Shigella dysenteriae* chromosomes.

Experiment E.4 Mapping by R-looping

At the correct temperature and in the presence of formamide, double-stranded DNA will denature to hybridize in preference with homologous RNA. Thus, a double-stranded DNA-RNA hybrid is formed, with the displaced single-stranded DNA forming the other side to the structure known as an R-loop (Fig. E.5).

Materials needed

Microcapillary tubes
  These should be repelcoated twice and washed well in distilled H_2O as follows: add one drop of Baycovin-Diethyl pyrocarbonate (BDH 44170) to boiling water and immerse the tubes. Rinse tubes thoroughly in boiling distilled water and dry

Cytochrome C
  Type VI from horse heart (Sigma 07752). Make up a stock solution of 1 mg ml^{-1}, filter and store in small amounts in the freezer. Bring to room temperature before use and discard remainder

Formamide (as in heteroduplex mapping)

R-loop buffer (1 M tricine-HaOH, 0·1 M EDTA, pH 8·0), millepore filtered

5 M NaCl

8·6 M urea

Procedure

1 Set up reaction mixture as follows:
  To 7 µl formamide add 1 µl of R-loop buffer, 1 µl 5 M NaCl and 1 µl containing 1 µg DNA/RNA
  Incubate in sealed microcapillary tube at 52°C for 16 h. Dilute further with R-loop buffer as necessary.

2 To prepare hyperphase add 55 µl formamide to 6 µl of reaction mixture diluted further in R-loop buffer to give an appropriate concentration of DNA and with marker DNA added.
  Add 30 µl 8·6 urea and make up to 100 µl with R-loop buffer. Heat at 40°C for 30 s. Place in ice-water and allow to reach room temperature. Immediately add 10 µl 1 mg ml^{-1} cyto-

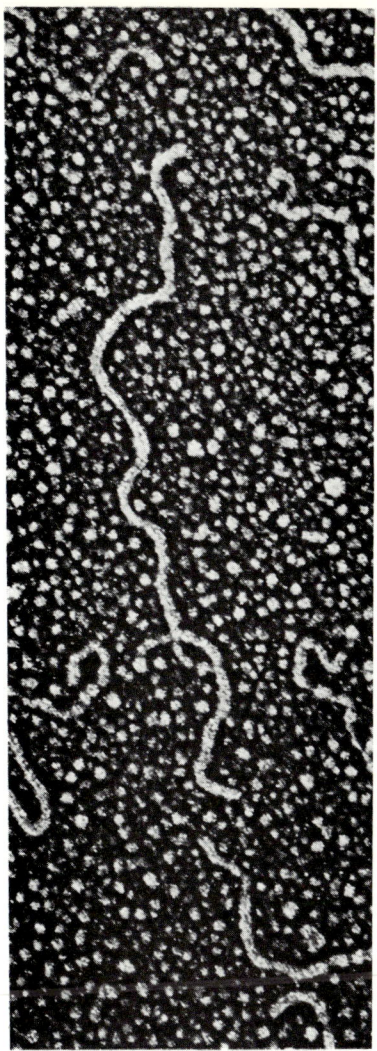

Fig. E.5 Hybridization of eukaryotic mRNA to a region of homologous genomic DNA forming an R-loop of 360 bases in a fragment of 4·5 kilobases. Polaron low angle, ion beam, gold shadowed. Dark field microscopy, Jeol 100 cx. A.V., 80 kV.

chrome C and spread on a deionized water hypophase. Stain and shadow grids as previously described.

References

Schafer, P., Boyd, C.D., Tolstoshev, P. & Crystal, R.H. (1980) Structural organisation of a 17 KB segment of the α2 collagen gene: evaluation by R loop mapping. *Nucl. Acid. Res.*, **8**, 2241.

Tilgmann, S.M., Tiemeier, D.C., Seidman, G.J., Peterlin, B.M., Sullivan, M., Maizel, J.V. & Leder, P. (1978) Intervening sequence of DNA identified in the structural portion of a mouse β-globin gene. *Proc. natl. Acad. Sci. U.S.A.*, **75**, 725.

Experiment E.5 Preparation of open circles from supercoiled DNA

The contour length of supercoiled DNA cannot be measured and so it is necessary to prepare an open circle form. This can be done simply by freeze/thawing which will nick the DNA and produce open circles.

Materials needed

Dry ice in ethanol
37°C waterbath

Procedure

Place the DNA in a small tube and freeze/thaw alternately five or six times, spread and examine. Open circle (relaxed) molecules can be readily distinguished from any supercoiled molecules which remain.

Index

Acrylamide reagents 130
Adenylic acid 10
Affinity chromatography 10–11
Agarose gels 101–2
   choice of concentration 101
   recovery of DNA 103–4
   re-use of 102
Alkaline hydrolysis 18
Anaemia, induction of 51
Aqueous droplet technique, for mounting double-stranded DNA 183
Autoradiographs, developing 135–6
Autoradiography 26
   gels 134–6
   use of fast tungstate screens 138

Bacteriophage λ, use as vector 145
Benton Davis screening 86
β globin genes, detection in DNA 80–90
Buffer B 126

Carbon-coated grids 176
cDNA
   annealing 32–3
   double-stranded
      homopolymeric tailing 27–31
      recovery of ethanol precipitates 23
      SI nuclease treatment 24–6
   recovery of ethanol precipitates 18–19
   single-stranded
      alkali treatment 18
      large scale synthesis 16–17, 18
      separation from unincorporated nucleotides 17–18
      small scale synthesis 17
   synthesis
      analysis of products 57–8
      first strand 15–19
      second strand 20–3
         alternative method 54
         large scale 21–2
         methods 20
         small scale 22–3
Cell lysis 43–4
Cells, competent 35–7
   factors influencing 59–60
Clones, storage of 60–1
Collodion-coated grids 175–6
Colony hybridization 39–40, 43–5
Cytochrome C 177

Dark field microscopy 180
DE 52 cellulose column 24
Denhardt's solution 85
Detergents, effects of 59–60
Dextran sulphate, and filter hybridization 84
Diethylpyrocarbonate 7, 8, 137
Difco methionine assay medium 120
Diphenyloxazole (PPO) 134–5

Disabled hosts, uses of 60
DNA
   annealing 32–3
   binding 43–4
   denaturation 43–4
   detection of β-globin genes in 80–9
   double-stranded
      homopolymeric tailing 27–31
      mounting 183
      recovery of ethanol precipitate 23
      S1 nuclease treatment 24–6
   electrophoretic separation of fragments 101–2
   extraction from mouse livers 166–7
   genomic, preparation of 157–8
   high molecular weight bacterial isolation 169
   hybrid, preparation of bank 155–60
   in vitro packaging 159–60
      preparation of extracts 161–4
      test extracts 163–4
   ligation of donor and vector 159
   packaging and size selection 148–9
   phage, replication 148
   plasmid
      preparation of 63–7
      transformation of *E. coli* cells by 35–8
   polypeptides synthesized from 122–5
   preparation of 4, 80
      and electrophoresis of samples, 100–1
   recovery from agarose gels 103–4
   recovery of ethanol precipitates of 18–19
   restriction endonuclease digestion of 80–1
   second strand synthesis 20–3
      alternative method 54
      large scale 21–2
      methods 20
      small scale 22–3
   single-stranded
      alkali treatment 18
      large scale synthesis 16–17, 18
      separation from unincorporated nucleotides 17–18
      small scale synthesis 17
      supercoiled, preparation of open circles 190
   synthesis
      analysis of products 57–8
      first-strand 15–19

Electroelution method of DNA recovery 103
Electron microscope
   preparation of grids 175–6
   use 179–81
Electron microscopy 171–90
   equipment and materials 175
   spreading nucleic acids 177–8
   staining and shadowing 178–9
Electrophoresis 26, 81–3
   of RNA 94–5
   onto DEAE-cellulose paper 103–4
   tank 101
   see also Gel electrophoresis

Escherichia coli
   CSH26 *rec*A   114
   CSR603   114
   transformation of cells   35–8
Exonuclease V   149

Filter hybridization, detection of β-globin genes   80–90
Fluorography, gels   134–6
Formamide   177
Freeze-thaw lysate (FTL)   162

Gel electrophoresis   57–8, 78–9
   apparatus for   96
Gels
   autoradiography   134–6
   fluorography   134–6
   preparation of   101
   SDS-polyacrylamide
      composition   131
      preparation   129–33
   see also Agarose gels
Gene expression systems   105–42
   comparison   108
Gene-bank, preparation   165
Glyoxal   94
Grunstein-Hogness screening   86

Haemin stock solution, preparation   52
Hershey medium   126
Hershey salts   126
Heteroduplex mapping   184–6
Holey carbon film   176
Hybridization
   box   84–6
   colony   39–40, 43–5
   filter   80–90
   probes   83
   washing DNA filters   87–8
   see also Northern blotting; Spot hybridization

K medium   126
Klenow enzyme   77

λ
   buffer   126
   genes
      arrangement   145–7
      expression in lytic cycle   147–8
   in vitro packaging   159–60
      preparation of extracts   161–4
      test extracts   163–4
   infection of ultraviolet irradiated cells   109–13
   phage repressor   145
   vectors
      choice   150–1
      choice of host strain   151–2
      cloning capacities   150
      preparation of gene-bank in   165
Low molecular weight mix (LMM)   139, 141
Lysate, freeze-thaw (FTL) preparation   162
Lytic cycle, expression of λ genes in   147–8

Mapping
   by R-looping   188–9
   heteroduplex   184–6
   restriction   72–9
   self-annealing molecules   187
Marker grids   41–2
Maxi-cell system   107, 108
Maxi-cells   114–17
Microscope slide agarose gel   49–50
Mini-cell system   107, 108
Mini-cells   118–21
   whole cell contamination of   121

NaOH method, for preparation of plasmid DNA   65–6
Nick translation   83–6
Northern blotting   93–8, 100
Nucleases, degradation of DNA   6–7
Nucleic acids
   electron microscopy   171–90
   length measurement of molecules   181–2
   spreading   177–8
   staining and shadowing   178–9
Nutrient broth   126

Pasteur pipette-sephadex column   17
Phages   109–11
   recombinant
      host strains   151–2
      maturation   149
Phaseolin   93
Photography   181
Plasmid pAT153   72
   mapping cleavage sites   74–9
Plasmid pBR322   72
Plasmid restriction maps   72–9
Plasmids   109
   analysis of   46–50
   rapid microscale preparation   47–8
   analysis   48–9
Polypeptides, synthesized from DNA   122–5

Recombinant clones, analysis of plasmids in   46–50
Restriction endonuclease
   digestion of DNA   80–1
   digestions   48
Restriction mapping   72–9
Reticulocyte lysate   12
   preparation and use   51–3
Reverse transcriptase   16
Ribonucleases, degradation of RNA   6
R-looping   188–9
RNA   3
   electrophoresis   94–6
   eukaryotic, preparation   93
   in vitro translation   12–13
   and isolation   6–9
   polyadenylated, purification of   10–11
Rotary shadowing   179

S30
   buffer, preparation   138–9
   extract, preparation   140–1

Scanning transmission electron microscopy
    (STEM)   180–1
SDS lysate, single colony   62
SDS-polyacrylamide gel electrophoresis   128–33
S1 nuclease   24
    characterizing preparation of   55–6
Single colony SDS lysate   62
Snap-backs   187
Sodium dodecylsulphate (SDS)   128
Sonic extract (SE), preparation   162–3
Southern blot analysis   90
    advantages and disadvantages   99–100
Spot hybridization   91–2

Transcription/translation system   107, 108
    preparation of components of   137–42

Transformants, recombinant plasmid analysis   39–45
Transformation   59–60
    frequency   38
    hosts, disabled   60
Triton lysis   63–5

Ultraviolet irradiated host system   107, 108, 109–13
Uranyl acetate stain   178–9

Vector
    insertion   148, 150
    replacement   148–9, 151